'Through *Gods in Shackles,* Sangita Iyer exemplifies the power we have to make change happen when we fully commit to acting on our passions with courage, commitment, and willingness to adapt in the face of adversity and challenge. Through natural storytelling and compelling writing, Sangita weaves a personal journey of courage and transformation. She invites the reader into her story and the stories of captive elephants, sharing her emotional connections, struggles, and triumphs along the way, revealing the deep disconnects we have in how we live our lives, and how we treat other sentient beings. Through this engaging story, Sangita Iyer helps us recognize how interdependence, community, diversity, and being open to adaptation and emergence create a transformative change for a sustainable future.'

—Dr Liza Ireland
Executive director of Changing Climates Consulting,
British Columbia, Canada

'Sangita Iyer's personal journey is a wonderful model for how we should, and must, "rewild" ourselves and learn to coexist peacefully with the fascinating, non-human animals with whom we share our magnificent planet. My hope is that self-centred humans will come to realize that the better we treat other animals, the better we will treat ourselves. It's a win-win for all. Some call the Anthropocene "the age of humanity". In reality, it is "the rage of inhumanity". And if we don't change our anthropocentric destructive ways, we will eventually live in an impoverished and very lonely world. Future generations will inherit the messes we leave them, and this surely isn't fair to them and to their children. *Gods in Shackles* is a courageous, riveting, and inspirational book that deserves a global audience, and I will do all I can to spread the word far and wide.'

—Marc Bekoff PhD
American ecologist, biologist, ethologist, and writer

'When a person stands up for injustice, the plight is infectious. Sangita's passion is infectious. The good people of this world came her way and helped because they believed in what she was doing. Sangita made *Gods in Shackles* and the VFAES happen, and bravo to her. I would thoroughly recommend this inspiring and uplifting book. The subject matter is brutal, but it gives me hope that humans can still make the world a better place if we change our ways and take a stand. Education and solidarity are the way forward. Excellent read and a fantastic journey!'

—Carla Kovach
Author of the bestselling Detective Gina Harte series

'Thought-provoking and very upsetting at times, particularly because that these tortures are under the semblance of religion. This wonderful book is deep and full of facts and emotions. I particularly resonated with the frequent parallels to the elephants' plight and Sangita's personal life experiences, both physical and mental. It seems incongruous and wrong that nature should bow to mankind, especially this most glorious and intelligent species—the majestic yet gentle elephant. Sangita is a determined, cause-driven, and passionate advocate whom I admire enormously. After a horrendous accident during which she was incapacitated for months, she becomes more and more determined and likens her suffering to those of the horribly abused elephants, particularly those in the temples. The book charts the making of her amazing film *Gods in Shackles*—an eye-opening, heartbreaking account of the barbaric cruelty inflicted on these naturally gentle giants to "break their spirit" and make them controllable to humans. I challenge anyone who has had the privilege to see these glorious animals in their natural habitat, in their female-orientated herds, to not be moved and become determined to support Sangita and her mission.'

—Rula Lenska
British actress and model

'Sangita Iyer possesses an abundance of courage, vigour, tenacity, and charm. It is the combination of these qualities which endures in the dynamic achievements of this outstanding author who is a dedicated campaigner for elephant protection. Sangita has demonstrated, time and again, that nothing stands in her way, dispelling untruths and sweeping away injustices so that egotistical and political barriers can crumble leading to change. This phenomenal, trailblazing book reminds us that humans need to self-heal in order to fully play our role in forming a synergistic coexistence with nature and animals. What we do to other species reverberates back to us. There is an urgency, therefore, to end the weeping of elephants and evolving of humans so that we all hold out hands with respect and tender love. Then elephants, and all animals, will be free from shackles and chains. In this outstanding book, Sangita Iyer is the worthy voice of elephants, the translator of their wisdom, of the mapping of their consciousness itself, and how it relates to us mere humans.'

—Margrit Coates
World-renowned author, healer
and interspecies communicator

'A painful and extensive injury becomes the occasion for Iyer's understanding of the horrific plight of the temple elephants. Her soul's understanding that the elephants are holy beings requires her to work on their behalf to free them from the literal ongoing torture of their ceremonial lives in captivity in the name of religion. The book weaves Iyer's story, the elephants' story, and India's extreme contradictions as modern life and its desperations devastate the ancient traditions.'

—Deena Metzger
Author of *A Rain of Night Birds* and co-editor of *Intimate Nature: The Bond Between Women and Animals*

'From an early age, Sangita felt a deep affinity with Asian elephants. Many years later, having long since moved to Canada, she became aware that the temple elephants used in religious ceremonies in her native India are horribly abused, even tortured. It became her mission to shine a light on this abuse. Here, she tells a heartfelt tale of creating and sharing the award-winning documentary film, *Gods in Shackles*, and draws parallels between her own inner search for spiritual freedom and her work to free her soul animals from the horrendous conditions of their captive lives. Along the way, Sangi shares with us glimpses of her beloved Indian culture and the wonder of the non-human world she finds even in the midst of cities. If you love our non-human brothers and sisters and find the Law of Attraction compelling, then this book is for you.'

—Dr Robert Kull

Author of *Solitude: Seeking Wisdom in Extremes—A Year Alone in the Patagonia Wilderness*

Gods in Shackles

What Elephants Can Teach Us About Empathy, Resilience and Freedom

Sangita Iyer

HAY HOUSE

Carlsbad, California • New York City
London • Sydney • New Delhi

Published in the United Kingdom by:
Hay House UK Ltd, The Sixth Floor, Watson House,
54 Baker Street, London W1U 7BU
Tel: +44 (0)20 3927 7290; Fax: +44 (0)20 3927 7291; www.hayhouse.co.uk

Published in the United States of America by:
Hay House Inc., PO Box 5100, Carlsbad, CA 92018-5100
Tel: (1) 760 431 7695 or (800) 654 5126
Fax: (1) 760 431 6948 or (800) 650 5115; www.hayhouse.com

Published in Australia by:
Hay House Australia Ltd, 18/36 Ralph St, Alexandria NSW 2015
Tel: (61) 2 9669 4299; Fax: (61) 2 9669 4144; www.hayhouse.com.au

Published in India by:
Hay House Publishers India, Muskaan Complex, Plot No.3, B-2,
Vasant Kunj, New Delhi 110 070
Tel: (91) 11 4176 1620; Fax: (91) 11 4176 1630; www.hayhouse.co.in

Cover design: Claudine Mansour
Jane Goodall photo: Stuart Clarke
Richard Louv photo: Used with Permission

A catalogue record for this book is available from the British Library.

Tradepaper ISBN: 978-1-78817-818-1
E-book ISBN: 978-1-4019-6885-4

Printed and bound by CPI Group (UK) Ltd, Croydon CR0 4YY

Dedication

This book is primarily dedicated to the gentle giants of our planet who are knocking on humanity's doorsteps, trying to awaken us—not to save their own species, but rather to prevent the collapse of the *human* civilization. I also dedicate this book to every single non-human, four-legged, feathered, and the tiniest of critters who appeared on my path to guide me. In addition, I honour those members of the human family who pushed me to delve deeper and connect with my shadow self, and those who drew out the very best in me.

Together, they enabled the writing of this book and gave me the license to shine a light on the myths and biases that are preventing human beings from conscious evolution. In this, I hope that my endeavours have served their small part in illuminating the interconnectedness of the sacred web of life.

Contents

Foreword

by Dr Jane Goodall
Founder of the Jane Goodall Institute and
UN Messenger of Peace

As I read *Gods in Shackles: What Elephants Can Teach Us About Empathy, Resilience, and Freedom* I was shocked, saddened, and angered by the cruelty towards the elephants who are forced to take part in religious ceremonies—cruelty that is described in this extraordinary book. I was amazed and moved by the courage shown by its author, Sangita Iyer. She loves elephants, yet despite the emotional pain she suffered when she saw the abuse meted out to them, she forced herself to visit as many of the temples as possible to record and expose their pain to the world. And when an accident left her crippled and in agonizing pain for months, she never gave up. Moreover, she realized that her pain, and the pain of the elephants, reflected the suffering of so many abused people around the world.

I have loved elephants since reading about them as a child. There was Tantor in *Tarzan of the Apes*—one of the few jungle animals outside of his ape family with whom Tarzan made friends and often walked with him in the moonlight or rode high on his head. And then I read *The Jungle Book* and met Hathi (named for *hāthī* (हाथी), the Hindi *word for* elephant)—the wise old bull, head of his herd. He was one of the oldest and most dignified of all animals in the jungle.

I have been fortunate, over the years, to spend a good deal of time watching elephants in East Africa, and whilst this book is about Asian elephants, the two species are very similar in their behaviour. Both Asian and African elephants have complex communication systems and are among the most intelligent of all animals, with big brains and long memories. Both form close-knit groups of related females and young, led by a matriarch. The males usually stay with their mothers for at least 10 years before wandering off to travel with other young bulls. And then, as they mature, the bulls move from herd to herd, seeking receptive females with whom they mate.

Elephants love to cool off in water, using their trunks to shower themselves. If possible, they like to submerge themselves so that only the tip of their trunks is visible. And they often plaster their bodies with mud from the riverbank before wandering off among the trees and resting in the shade. They try to avoid staying out in the open during the heat of the day. Elephants have very distinct personalities. They are caring and compassionate and have often been observed gathering around distressed or sick individuals, touching them with their trunks and making soft reassuring calls, clearly trying to comfort them. Like us, elephants know joy and sadness, fear and depression. Like us, they feel pain.

African and Asian elephants are threatened with extinction. Habitat loss forces them into contact with people and conflicts arise when elephants raid the precious crops of village farmers. Both elephants and humans may be killed or injured in conflicts of this sort. African elephants are ruthlessly hunted and killed for their ivory; Asian elephants are taken into captivity, to be exhibited in zoos around the world, trained to perform in circuses and made to carry tourists on their backs.

And, as described in distressing detail in this book, elephants are forced to play a major role in religious ceremonies in Asian temples. Most of the Temple elephants are captured as youngsters from the wild, torn from the love of their mother and families. These innocent, fun-loving, and highly emotional youngsters must then go through a period of brutal training where they are shackled and beaten until their spirit is broken, and they will instantly obey the commands of their mahouts.

This is all the more shocking because in both Buddhism and Hinduism the elephant is revered. The Hindu God of Beginnings is Ganesha or Ganesh, depicted with the head of an elephant. In Buddhism, the elephant stands for strength, honour, patience, peacefulness, and wisdom. Both these religions teach respect for animals.

To continue to capture or kill elephants in the wild is leading to their extinction, to the detriment of the habitats where they live. To continue to exploit them in captivity, whether wild caught or captive born, is to perpetrate unacceptable cruelty to highly intelligent social and sentient beings. Of course, those whose livelihoods depend on the exploitation of elephants, the mahouts and the owners, must be helped to find other ways of making a living.

My first meeting with elephants in Asia was when I visited a facility in Nepal that was doing its best to give animals rescued from slavery a better life. They were trying to raise money to provide a large area where the elephants could roam freely, but in the meantime, although they did get out each day to browse in the forest, the bulls were still shackled. And as I watched two of them hobbling forward, their front legs tightly chained together, it broke my heart, and my admiration for Sangita increased. For she, so passionate about elephants, forced herself to witness their suffering, hiding her broken heart as she got permission to film what is going on—how they are forced to stand for hours on hot concrete under the fierce midday sun, beaten, jabbed in their sensitive areas with bull hooks so that many have severe and untreated injuries. She did this to raise awareness of the plight of these shackled elephants in order to bring in legislation to forbid the cruelty and to educate people so that they would speak out against it.

The public, especially children, must be better educated about elephants and their needs in particular animals in general. Let us work together to bring this about so that no elephant, ever again, will be abused, enslaved, or shackled. Future generations of elephants must not experience captivity; they should be protected in their natural habitat or the comparative freedom of really good sanctuaries. I pray we all do our part to ensure this vision comes true.

Introduction

A Shared Song

We live in a time of extinction and resurrection. In coming years, without our action, wildlife will be radically diminished in number and diversity. But we may yet turn the corner, away from the Anthropocene—the age of man and loneliness—toward an age of symbiosis and joy.

Sangita Iyer is one of the people helping us step back from the path towards an abyss. In *Gods in Shackles*, she personifies that choice. She discovers the ecstatic moment when despair—both personal and global—gives way to sorrowful hope. She does this in the company of elephants, urging us to save them, and to save ourselves, to return to the family of animals. She describes her own struggle with loss and loneliness, a story to which many readers will personally relate. Medically, human isolation now equals obesity as a cause of early death, not only because of suicide but because of the diseases associated with loneliness. This is a crisis of physical psychological and spiritual health, and it is rooted in our loneliness as a species.

We are desperate to feel that we are not alone in the universe. We are not. A whisper is all around us, the constant song of life communicating with itself. In *Gods in Shackles*, Iyer links our own resurrection as individuals and as a species to this shared song.

The fate of elephants and humans is down a road we both travel. As she reports, between May and June 2020, approximately thirty-three elephants were massacred within a span of forty days, directly related to the Covid-19 crisis as it unfolded in India. 'Many of them poisoned, some were killed using explosives, some electrocuted, and some shot to death for their precious tusks to be sold in the illicit ivory trade market,' she writes. Elephants and people live on the same landscape, but humans have not learned how to share their world. The pandemic of 2020, and it will not likely be the last, grew out of human cruelty to animals. It has shown the *necessity* for coexistence in coming decades, as people and wild animals increasingly share the same space.

The crisis concerns our health, but it is not just about us. Animals have a right to exist even if we didn't benefit from them. In environmental ethics, this is called 'existence value'. Still, human culture must acknowledge that our health depends on the well-being of other species and the health of the Earth itself. In the public health field, this mutuality of interest is expressed in the One Health movement, which holds that none of these elements can be held separate from the others. All must be treated; all must be nurtured.

Our species will not survive in an age of loneliness and human narcissism. It will thrive only through the principle of reciprocity. We must care about the Earth and the children of all species as much as we care about our own.

Gods in Shackles suggests this question: What will it take to move our species on climate disruption, biodiversity collapse, the threat of mass extinctions, and the spread of human pathologies? Facts, science, logic are essential, but as we have learned in recent years, data is clearly not enough to sway hearts. Those of us who care about the future need at least two additional ways to make the case. The first is love—deep emotional attachment to the nature around us. For Sangita, that love is a focused light, shining on the elephants she hopes to save. The Australian eco-philosopher Glenn Albrecht argues that only 'a shift in the baseline of emotions and values has worked' to transform facts into action in other areas such as feminism, marriage equality, and racial justice. The reason these unfinished causes have made progress, Albrecht argues, is because 'they revolved around the issue of love'. This is why images of suffering animals and heroic stories of human beings working to save them are so important. They remind us, at least for a while, that we belong to a larger family, one worth loving.

The second element is imaginative hope—our ability to describe a future worth creating. In these pages, Sangita Iyer offers us both love and imaginative hope. Hope becomes more realistic when we view the four horsemen of the apocalypse—climate disruption, biodiversity collapse, extinction, and the decay of human hope—as a single existential threat with shared solutions. To find that path, to take action, we must first listen to the song that surrounds us, as Sangita has done in her own life.

—Richard Louv
Author of *Our Wild Calling* and *Last Child in the Woods*

Chapter 1

In a Split Second

It was 9 January 2017, around 11:00 a.m., and I was being rushed to a hospital in an olive green jeep, my ruined foot turning a disturbing shade of dark blue as I watched. An elephant had knocked me down, causing five broken bones in my ankle and pitching my carefully planned trip into chaos in one horrible instant. The driver, with a thick black moustache, clad in khaki uniform was manoeuvring the vehicle like a maniac, stirring up dense clouds of mud from the unpaved roads. It would take us at least two hours to get to the nearest hospital. There were none in this remote area of Kappukadu in the southern Indian state of Kerala.

The wildlife warden of Trivandrum district, Shaji, was sitting in the passenger's seat, sweltering and frantically making phone calls to rearrange his day. My accident had occurred just as he was leaving the Kottoor Elephant Rehabilitation Centre, a government-run forest camp, for meetings in town. Over the phone, he informed his boss about what had happened, the tension in his voice and on his face was palpable!

In the back seat, I was lying in the most uncomfortable position, my head resting in the arm of a young tour guide, Sajna. She was seated on my right, caressing my sweaty forehead, and trying to get me to relax. My legs were stretched out on the lap of a middle-aged woman, Sita, who was a cleaning lady at the forest camp. She delicately held my swollen left foot that had collapsed on one side, clinging on to the few remaining ligaments of my ankle. Sita was right there with me when the accident had happened and became so distraught that she insisted on accompanying us to the hospital.

Sajna spoke to me in a calming voice. 'Everything will be okay, ma'am,' she said softly, while Sita stroked my swollen foot. She was trying to soothe my agonizing pain, which intensified each time the jeep jolted over a bump in the road.

The air conditioner was on full blast, yet its output of cool air was meagre. I asked that the vehicle's windows be rolled down. As they did, a wave of hot, humid air filled the jeep instantly, and so did the obnoxious sounds of trucks and cars. Traditional songs blared through large stereos as hawkers on the sidewalks were yelling at the top of their lungs, trying to sell fruits and vegetables. I could even hear cows mooing and rickshaws speeding through the narrow street. Unable to handle the chaos, I asked the driver to roll up the window as I tried to relax.

Every now and then, I asked how long it would take to get to the hospital and the driver's response remained the same, 'Another fifteen minutes.' Two hours later, we were on the outskirts of Kerala's capital city of Trivandrum. Here, we found a different kind of chaos. Bumper-to-bumper traffic, intense air pollution, and a louder and more obnoxious cacophony of vehicles honking. Dark clouds of toxic soot spewed from the exhaust pumps of massive trucks and buses. I began to feel dizzy and forced myself to shut my eyes to escape all of it.

It took us over thirty minutes from the city outskirts to arrive at the Trivandrum Medical College Hospital. We drove towards the white building through an arched, black gate and made our way to the emergency unit. As we did so, I realized that the facility was run by the government. Hundreds of patients were packed at the main entrance for a preliminary assessment, indicating that, perhaps, there was a significant shortage of rooms and/or medical personnel. But fortunately, the forest officials managed to pull some strings and someone from the hospital staff promptly tended to me.

A man dressed in a light blue uniform brought a stretcher to the side of the jeep. I painstakingly sat up, struggling to remove my foot from Sita's lap. Somehow, I lifted myself from the seat and was devastated to find that my foot had swelled dramatically during the two-hour drive. An excruciating pain shot up through the left side of my body as I tried to lower my foot. It felt like a heavy rock was being pulled down through my leg by the earth's gravity. Meanwhile, the officer and Sajna grasped each of my arms and Sita gingerly held my injured foot as I lifted myself from the car. Balancing my body on my right leg, I managed to lie on the stretcher.

Inside the emergency area, hundreds of patients were sitting in wheelchairs or lying on stretchers barely a foot away from each other. Some coughed and sneezed without covering their mouth. Being a germaphobe, I cringed listening to the hoarse sounds of sick patients. Communicable diseases such as hepatitis, tuberculosis, malaria, yellow fever, and dengue fever continue to plague India, despite all the medical advancements. I gripped my nostrils, trying not to breathe, but couldn't hold it for too long.

My constant comfort, the two women, stood on each side of my stretcher. One rubbed my forehead while the other offered me a bottle of water. As the driver and Shaji were talking to a man in a white apron, a tall dark man with a thick black moustache joined in. It was the friendly forest ranger, Sukesan, whom I'd met at the Kottoor Elephant Rehabilitation Centre. Apparently, he had driven all the way to make sure that everything flowed smoothly. A few minutes later, four men approached me and handed over a few application forms to fill in. Soon after, they wheeled me into a cubicle for a blood test. Subsequently, they took me to another area for a number of X-rays and other routine medical tests. They then snaked their way through dark hallways and pushed my stretcher into a tiny room in a corner. It felt like a prison, and I began to hyperventilate.

My mind suddenly drifted back to a damp and dark dungeon near a temple where I had seen a lone elephant tethered cruelly with short chains. He was standing on his urine and excrement, swaying back and forth, bored out of his mind. In his trunk, he held a bunch of dry palm leaves, using them to dust his feet and shoo away the bugs. I also noticed a ring of light pigments around his ankles and a heavy chain tossed over his body.

My thoughts were abruptly yanked back into the dim enclosure as two young men in white aprons pushed open the door and entered. One of them—an orthopaedic specialist—began to interrogate me. The other one grabbed my twisted ankle and mercilessly rotated it back into its position. Unable to bear the persecution, every cell of my being screamed in agony and my shrieks reverberated through the hallways. He then put on a temporary cast to hold together my leg and foot.

The surgeon later explained that as agonizing as it was, he had to spin my foot back into its original position in order to avoid future complications. There was a short silence and then he broke the bad news. I had sustained multiple fractures on my left ankle and foot, and I would need a surgery. It felt as though a bolt of lightning had struck, triggering a downpour of tears. The two surgeons, seemingly indifferent, walked out of my room, leaving me alone to deal with my misery.

The team of four people from the forest department, including the two women, walked in hurriedly and wheeled me out of the room. And then, cutting through a packed hallway, they took me into a stuffy elevator that emanated a potent and musty smell. As it suddenly ascended, a queasy sensation exploded in my belly, exacerbating my anxiety. Scanning around for Sajna and Sita, I found my constant companions who had never left my side since the accident had occurred. They were standing behind the stretcher right next to my shoulders. Sajna placed her soft palm on my forehead and said, 'Don't worry. We are here, ma'am.'

Sukesan was waiting on the third floor near the elevator door when we disembarked. His usually cheerful face looked droopy and disappointed. 'What's the matter?' I asked. With a heavy voice, he replied, 'Madam, the operating theatres here are booked for the next three days. You've been placed on the waiting list.'

My heart sank. There was absolutely no way I could wait around for three days in a congested government hospital. There was too much to do! I was in Kerala on a mission to end the atrocities against the elephants, and I had scheduled meetings with some of the top government officials in the state. A wave of panic swept through my body, making me feel sweaty and claustrophobic. Distressing as it was, I closed my eyes, trying to breathe deeply and inhaled every ounce of air that my lungs could take in. I laid there helplessly, praying fervently for some kind of divine intervention or a miracle.

Suddenly, someone gripped my right shoulder abruptly. It was Sukesan. With a worried look on his face, he asked, 'Are you okay, ma'am?' I simply nodded my head, but he wasn't convinced. Then, a light bulb suddenly went off in my head. I picked up my cell phone and began to call some influential people I had met during my previous visits to Kerala. Most calls went unanswered, but it took just one person to make all the difference. That one person turned out to be a prominent woman who (asked to remain anonymous) promptly answered the phone. When I explained to her what had happened, she asked me to call her back in ten minutes.

Meanwhile, I frantically phoned my family in Mumbai, Ooty, and North America. I could sense the shock and sadness in their silence on the other end after they heard about the accident. I felt homesick and desperately wanted to return to Toronto. But that was not an option.

Half an hour had gone by, and I had not heard back from that prominent woman. I rang her again. She informed that she had made arrangements for me to be admitted in a private orthopaedic clinic, the famous SP Fort Hospital. So, they wheeled my stretcher back into the same stuffy elevator and into the emergency area.

As we were waiting for the ambulance, I noticed a group of familiar faces chatting away. Ashwathy, the concierge manager from the hotel I was staying at, had arrived at the hospital to check up on me. She was accompanied by a few other hotel staffers. When she saw the condition I was in, her jaw dropped, and her eyes widened. She could not conceal her shock. But she assured me that the staff would do everything possible to make my stay at the hotel comfortable after my surgery.

By now the ambulance had arrived. My stretcher was wheeled out of the emergency waiting area and placed as close to the vehicle as possible. I managed to stand on my right leg, with my left foot folded behind, as it was throbbing in agony. Somehow, the two women from the forest department helped me into the back seat of the four-wheeler.

By the time we left, it was around 5:00 p.m.—peak traffic time. Lying in an ambulance with red lights flashing and sirens screaming, felt like a scene out of a movie. The vehicles around us moved in incremental fits and starts. As a result, it took us at least thirty minutes to get through the congested city roads to the SP Fort, just a couple of kilometres away.

I was then transferred to another stretcher and wheeled into the critical care unit. I suddenly felt my heavy bladder and realized that I hadn't used the bathroom since morning. But I couldn't walk! I had to depend on a nurse and a bedpan. My 'independent' life was beginning to disintegrate.

At around 8:00 p.m., I was wheeled in my stretcher to the fourth floor. Up until this time, Sajna, my tour-guide friend, never left my side. Her presence was deeply comforting. But here on, I had to be on my own. At around 8:30 p.m., I was taken to a dim, sterilized room where the only person allowed was the patient.

The potent smell of antiseptic agents filled the air and stung my nostrils. A nurse who was working away on her computer casually walked me through my upcoming surgical procedure. Soon after, an anaesthesiologist arrived and wheeled me into the adjacent operating theatre. Inside, at least six people, including two female nurses, all adorned with capes, gloves, and masks, were in attendance to assist with the surgery. As the anaesthesiologist was searching for a specific vein in my spine to tranquillize me, the top orthopaedic surgeon in Kerala, Dr Cherian Thomas, walked in. He had been called into the hospital late in the day to conduct my surgery. Dr Thomas cheerfully shook hands and assured me that everything was going to be just fine. I smiled and nodded my head, but my trepidation was all too obvious.

I was about to go under the knife for the first time in my life! I laid there helplessly, in a strange place, with no family or friends for support. I had no choice, but to surrender—an act of submission that I would replicate many times in the months and years ahead.

Dr Thomas walked away from me, signalling the nurses. Soon, a screen was placed between my torso and hips, covering the lower portion of my body. It served to obscure my view of the surgical tools, which would have most likely terrified me. As the surgeon began to examine my left foot, the anaesthesiologist inserted the needle into my spine. It was a tiny poke, and in a few moments, I was out like a light. The last thing I remember was Dr Thomas lifting my injured leg and releasing it, allowing gravity to take care of the rest.

Early next morning, I was awakened by the sun's glorious rays in the hospital room. Sita, my faithful shadow, was wide awake, resting on a narrow bed right across from mine. Outside, the red ball of fire was rising behind some palm trees. Two eagles flew past the golden rays, creating a picture-perfect silhouette.

My window overlooked the Padmanabhaswamy Temple, the world's richest temple. The brilliant sun beamed on its golden roof, shaped like an open umbrella. I had been to Trivandrum many times but had never visited this popular temple. Here it was, right in front of me, as though reminding me to stop over at the home of my Hindu god, Lord Vishnu.

Out of nowhere, an intense pain shot from my left foot. I looked down, and there it was, elevated on two pillows, wrapped in a white cast with layers of stiff bandages. The hard, outer layer was made of plaster of Paris, the same material I had used for my art projects in school. As I was trying to get used to my new reality, I noticed a catheter connected to me. It was all too much to handle. I took a deep breath and closed my eyes.

It was beginning to sink in that the accident was more serious than I had imagined. When Sita saw my incredulous expression, she explained that it was a three-hour surgery. I was wheeled into this room in the wee hours and transferred onto the bed where I was now lying. She told me that I would remain here for the next five days and assured me that she would take care of me for as long as I needed.

My heart began to overflow with grief. Sita placed her palm on my forehead and as she was consoling me, there was a knock at the door. It was Sukesan, the forest ranger. Apparently, he had stayed overnight in the hospital, sleeping on a couch near the reception area. His and Sita's dedication touched me profoundly.

Meantime, a loud roaring laughter filled the hallway. I quickly wiped away my tears, trying to put on a brave face. Dr Thomas was making his morning rounds, along with four other doctors, including a physiotherapist named Dr Padmanabhan. As the doctors converged, Sukesan promptly left my room. After examining my foot, Dr Thomas pulled out the X-rays that had been taken after my surgery.

The devastating images revealed two titanium plates

and eighteen screws holding together my left leg and foot. In a matter-of-fact tone, Dr Thomas explained that I had sustained five broken bones. Apparently, it would be at least twelve weeks before I could even place my foot on the ground. I looked at him in utter disbelief. He squeezed my left shoulder in support, and, in an optimistic tone, said that I would soon be running. I knew that was far from the truth.

Moments after the doctors left, a nurse showed up to give me a sponge bath. Following this, I was permitted to discard my hospital attire and put on street clothes. The nurse then dragged an IV station to my side and proceeded to poke me with the needle to inject the drip, while also injecting a shot of painkiller on my upper arm.

As the nurse left, Sukesan returned to my room. Laughing uncontrollably, he handed me a major newspaper. But what I read wasn't funny at all. On the contrary, it was distressing. My accident had made the headlines—it was a sensational story, factually inaccurate and totally exaggerated. Sukesan explained that my accident had been covered by all the state newspapers and major TV networks. He also added that the forest officials wanted to ensure that this unfortunate episode received broad media coverage to avoid the spread of any misinformation about the government-run elephant camp. What he didn't realize was that in order to cover up some of their own shortcomings, they had put the entire blame on me for the mishap.

I used to cover such tragic stories as a television journalist just a few years back. And now, to find the proverbial shoe on the other foot was a bit ironic, to say the least. It felt terrible to be the subject of the story, especially one that had been distorted. My phone rang. It was one of my journalist friends in Kerala who wanted a quote from me regarding the accident. I pleaded with him to omit the story from his newspaper and he obliged. I turned to Sukesan after noticing a smirk on his face. He understood the irony of my situation and apparently found it amusing that the media was hounding me.

After chatting for a few minutes, Sukesan left the room. As I retreated into my bed, it dawned on me that I was facing the most serious health crisis of my life. I wondered if I would ever walk again. However, I only had a few moments to indulge in hopelessness and misery before the door flung open again. I was already exhausted by the onslaught of visitors and wanted to be left alone for a few hours to comprehend what had happened to me.

But as I turned to the door with an irritated look, the sweet smile of my young nephew warmed my heart. Abhi announced that he had taken a week off from his brand-new job and flown in from Mumbai to spend a few days with me. The next day brought me another sweet surprise—my brother, Raju, showed up. Growing up in Mumbai, we'd had a difficult childhood. Although we fought frequently, we also loved each other deeply. It was comforting to have my brother by my side. As soon as he looked at my foot, his face turned grim. He became utterly speechless, trying to grapple with the seriousness of my fracture and its magnitude.

Just two months ago, Raju had joined me in Goa where my epic elephant documentary, *Gods in Shackles*, was screened at the prestigious International Film Festival of India (IFFI). He and his friends had endured a gruelling eleven-hour drive (approximately six hundred kilometres) from Mumbai to attend the award ceremony that lasted only a few hours. He had been blown away by the film and the red-carpet treatment that it had received. Soon after that, I travelled to Sri Lanka for twenty days of screenings before returning to Mumbai. During our subsequent reunion of sorts, he and I cherished the time spent with my mother, catching up on life.

My family had never seen me sick or helpless. In fact, I had earned a notorious reputation of being too independent a girl for a Brahmin family. Some of my relatives thought I was inviting death with my fearless journalism. Others looked down on me as a nomadic globetrotter who cared only about animals.

Although my parents were traditional in many ways, they were also a bit unconventional. For instance, they didn't have an arranged marriage, and both of them were employed. It was a big deal for women to be working in India in the seventies. My mother was well educated and had reached the status of an income tax officer in short order, while my father served in the Indian army. My parents worked very hard so that their children could receive the highest education and there would be ample money for my dowry when they got me married.

The expectation, of course, was that I would wed and start a family. But at that point in my young life, I was more intent on developing a career in the media industry. Although I loved the natural world and life sciences, I wanted to become a TV reporter. But my parents wanted me to become a doctor and so I ended up obtaining a bachelor's degree in biology. I soon realized that dissecting insects and frogs wasn't for me, let alone cutting up a human body. Thus, I pursued botany and ecology and went on to teach natural sciences in a primary school.

While I was relatively content professionally, there was a profound sense of dissatisfaction. I was still living with my parents. They controlled everything I did, and I began to feel like a bird in a cage. For instance, I enjoyed wearing Western clothes and cosmetics. But my parents insisted that I wear only traditional Indian outfits, and I was chastised when I wore make-up. I wanted to learn guitar, but my parents sent me to learn classical music.

I longed to explore a whole new world and be free to do what I chose, when I chose, and with whom I chose. Sick and tired of feeling confined, I moved to Canada in 1989. I underwent culture shock, seasonal shock, career shock, and language shock, among other things, all of which turned my entire life upside down. But I was never afraid of hard work and self-improvement. I took up speech therapy to soften my British Indian accent so that people could better understand me. I was also very ambitious. I worked a full-time and three part-time jobs. One of them entailed selling alarm systems, knocking on doors during bone-chilling Canadian winters. The blistering winds made me feel as though my frozen nose would chip off and I'd never be able to breathe again.

During my early years in Canada, I pursued my dream of becoming a broadcast journalist by returning to school for postgraduate journalism studies. And although most people felt that the odds were very low, I ended up becoming a videographer in one of the most prominent multicultural television networks in Canada. Despite the fact that biology and ecology were my specialities, I had to settle for a general reporter's job after I graduated because environmental stories seemed irrelevant unless they were controversial.

However, after a five-year search, I landed my dream job at Bermuda's ABC/CBS affiliate, which also gave me the opportunity to independently produce a thirteen-part series called *Bermuda Shorts*. The success of these short films catapulted the creation of my own charity, Bermuda Environmental Alliance (BEA), which was another dream come true. This allowed me to produce a six-part documentary series entitled *Bermuda–Nature's Jewel* which focused on Bermuda's diverse and pristine ecosystems. It was featured on Discovery Channel Canada and Bermuda's local television stations and is currently being used in the island's school system. My position with the network lasted until September 2008. After it ended, I travelled back and forth between Toronto and Bermuda, managing and directing the BEA until February 2014.

Just as I was wondering what the next chapter of my life would entail, I went through a series of fateful encounters that I detail in this book. I was led to cross paths with the elephants of India, my soul animals. Captive elephants, in particular, evoked painful childhood memories and emotions that were deeply buried within me. The profound insights that they offered would inspire me to direct and produce the multiple award-winning documentary film about their sad plight. This was *Gods in Shackles*—which I was in the middle of promoting when the accident occurred.

In all of this, I was enjoying the very pinnacle of success. And yet here I was, lying helplessly, unable to walk and tend to my basic needs. On top of this, I had become so dependent on other people that I stopped trying to do things on my own. The independent woman that I had once been was now a pathetic and useless shell of her former self. All I could do was eat, lay on my hospital bed, and look out of the glass windows. What a catastrophe!

As I struggled to pick myself up and sit on my bed. I glanced at Raju. And judging by the expression on his face, it seemed that he was still gripped by shock and disbelief. After a few moments, he mustered up the courage to ask what had happened. As I began to narrate the story, I remembered that Sajna had videotaped the entire ordeal on my iPhone. So, I hit the play button and handed the phone to my brother.

For the first few minutes, Raju and his son laughed uncontrollably, but suddenly their expressions turned intense and serious. Watching them, my body stiffened, and my foot became extremely tight. I could hear echoes of people yelling and screaming and the haunting sounds of shackles and merciless beatings. My heart raced as I relived the terrifying drama that had unfolded just twenty-four hours prior. Choking back tears, I pleaded, 'It wasn't his fault; it was all mine. Don't get upset with him!'

The next few moments were filled with a heavy silence. Then there was a knock at the door. It was Ashwathy, the concierge manager at the hotel that I was booked into and where my brother and his son were staying for the week. She had brought me undergarments and laundered clothes. She graciously offered my brother and his son a ride back to the hotel. It was good to know that Raju would visit me again the next day and then intermittently for the next several days.

I looked outside to see that the skies were lit in yellow, amber, and orange colours. Dusk was settling in and the birds were returning home. A flock of ravens descended on the palm tree outside my window as elegant white cranes flew past. Suddenly, again, I felt like a confined prisoner. I took a deep breath and closed my eyes, unable to bear the thought of being so restricted for the next several months. My foot began to pulsate in pain, and I cried myself to sleep.

The next day, I had to go through the same routine: a sponge bath, a shot of painkillers, and then the drips. By the third day, I had been weaned off from the fluids. Dr Thomas and his team brought in a walker and taught me how to walk with it. They asked me to balance my right foot gently without straining my arms too much. Then, with a stern look on his face, Dr Thomas said in a commanding voice that under no circumstances should I rest my left foot on the ground.

During the next five days, there was never a dull moment. Top officials from the forest department dropped in unannounced. It included the—now retired—additional secretary of the forest department, Mara Pandian, the chief government veterinarian, Dr Jaya Kumar, from the Kottoor camp, and even the—now retired—chief wildlife warden with whom I'd had a strained relationship in the past. Apart from it, Kerala's police chief, Lokanath Behera, phoned to check on my progress and assured me that he and his wife would visit me in my hotel after my release from the hospital.

Five days later, it was time to say goodbye to Sita, upon whom I had become quite dependent. So unconditional was her service that she rejected even a small token of appreciation. Her eyes brimmed with tears as we hugged each other. It was hard to say goodbye to such a sweet soul.

Ashwathy and her team from the hospital, together with my brother and nephew, came to collect me and my belongings. After settling the bill, the hospital staff wheeled me down to the cab in a wheelchair. It was truly comforting to be surrounded by so many loving and caring people. It was now time to heal in solitude.

My leg, which had walked bravely to places that most people would dare not venture, had to mend. It had served me unconditionally all my life, and now, it was my turn to repay the favour by giving it some rest. Perhaps I had taken my foot for granted. The old clichés are true. We tend to appreciate the value of something only when we lose it, even if temporarily. When we find it again, we cherish it more than ever before. Eventually, if we're mindful enough, we learn to value everything we have. This would be the first of many lessons that I would learn on my journey to help the elephants of India.

As I waited by the hospital reception for the hotel van, I couldn't help but wonder why total strangers did such kind things for me. Sukesan didn't have to spend two days and nights around the clock at the hospital. And why had Sajna and Sita insisted on taking care of me when they had their own families to care for? None of them were forced to make the sacrifices they had made on my behalf.

But from the moment the accident had occurred at the elephant rehab camp, these individuals never left my side. All the barriers—language, caste, creed, and religion—had disintegrated. There was no anger, resentment, blaming, or shaming. The entire forest department had come together to support me. They tolerated my impatience and frustration. Somehow, these amazing people found a way to look beyond my frailties and offered unconditional love. They showered heaps of praise on me for what they considered as sacrifices I had made to serve the natural treasures of our earth—the elephants. They expressed their true nature by allowing their innate qualities of love, compassion, and kindness to shine through. I had experienced humanity at its best.

Initially, after the accident, I had a difficult time trusting total strangers. But when my pain and suffering became unbearable, I had no choice but to let go of my brittle anxiety and embrace them. Somehow, my vulnerability taught me to trust, and in trusting, I realized that people are intrinsically good.

For me, this was a rich, growing experience. Through these trying times, I turned to my inner being for some solace, trying to find something positive out of the catastrophic accident. I understood that while physical pain and suffering are temporary, the impressions we create in the hearts and minds of others remain permanent. This lesson in trust would ripen and deepen within me. It was one that I would draw upon many times to face the myriad challenges in my path.

Indeed, in a split second on that fateful day, my life had changed forever on so many levels. Little did I know that the accident would catalyze my inner growth and reveal profound insights pertaining to my unfolding destiny with the abused temple elephants of Kerala, where I was born and raised.

Chapter 2

Learning to Heal

My release from the hospital meant that I no longer had to feel like a patient and could continue to recuperate at my hotel. After five relentless days of visitors at the hospital, I could now reclaim my personal space. It wasn't much later that Ashwathy took me to a room designed for people with special needs. As I walked into its large bathroom, I noticed that there were handles in every nook and cranny. I shut the bathroom door and broke down as another jolt of reality sank in. In addition to my walker, I now needed handles for support. I was disabled, even if temporarily.

All my life, I had enjoyed perfect health. A vegan by choice, I thrived on grains, fruits, and vegetables, much like the elephants did. I refused to take medication even for migraines and the common cold. Instead, I turned to yoga for the dis-ease to vanish. I had promised myself that chemicals would not contaminate my body.

But here I was, trying to deal with my new, harsh reality. Foreign objects—two titanium metal plates and eighteen screws—were inside my body. I needed them to hold my feet and leg together so I could walk, even with a walker. I had also become a perennial pill popper since my surgery in order to mask my chronic pain and to avert bone problems. It was just a year prior that my orthopaedic specialist had said that at fifty-three years of age, my bone density was that of a thirty-five-year-old. I had chuckled and replied that perhaps my walks in the wilderness were the reason for my good bone health, and she concurred.

But now, I couldn't walk in the wilderness to listen to the soothing melodies of robins and cardinals. I couldn't run up and down the steps in the woods. I couldn't even walk without the aid of a walker. Depressed, I looked at myself in the bathroom mirror. The white plaster-of-Paris cast covering my left leg, beneath my knee, seemed like a lifeless object dangling. I felt totally disconnected from the alien entity attached to my left side.

Then I heard a knock on my bathroom door. My brother said they were getting concerned. I realized that I'd been in the bathroom for twenty minutes and had forgotten that everyone was waiting for me. I wiped off my tears and entered the room. It was a fairly large one and had a small table where I could use my laptop. It also had two comfortable, upholstered chairs and a large TV. Half of the wall facing the main road was covered with large glass panes to allow in natural light, but there were no windows. It was a self-contained room and had everything I needed.

Later that afternoon, Aathira from reception phoned to inform that she was on her way to see me. A tall, slender 19-year-old young woman walked into my room with a gorgeous smile. Her dark brown eyes were filled with excitement. She declared that she would be my primary caretaker, and, in her absence, a 23-year-old woman, Neethu, would look after me. I had met Neethu before, and she seemed mature, thorough, and meticulous.

Aathira was now going to bathe and groom me, my first real bath in nearly a week. She first brought in two buckets of hot water and then mixed it with cold tap water. She also had wrapped my fractured leg in a black garbage bag to prevent water from seeping into the plaster. It was now time to remove my clothes. But it was too embarrassing. Tension filled the air. After a few moments of hesitation, I mustered up my inner strength to remove my nightgown, by holding the walker, and then sat on the stool.

In her childish manner, Aathira began to pour water on my back. In that very instant, the knots around my neck melted and my entire body began to relax. It was the same soap I had been using for years. But that day, the fragrance was particularly enchanting. After the delightful bath, she massaged my back, neck, and shoulders and helped me into my clothes. She then made me a cup of coffee before she left.

I could not wrap my head around how helpless I had become, for I had always been independent. This was the beginning of disintegrating my rather well-defended walls and allowing others into my secluded life. Perhaps this was the universe's way of breaking down my tough outer shell and helping me realize that we are all truly connected.

Over the next few weeks, bathing became a sacred ritual, and being around these young women calmed my mind. I could depend on them to buy me fruit, cut my toenails, massage my back, and bathe me. I could even share my trepidations with them. They often reminded me that I was like their mother, and we developed a special bond. We had become so intimate that they would visit me even on their off days. They showered unconditional love, mercy, kindness, compassion, and service.

That evening, my brother dropped in briefly to organize my clothes and then returned to his room. I had no visitors scheduled for the upcoming week. Finally, I could be alone. After five days of intense medical attention and non-stop visitations, I cherished my solitude. I could now try to digest what had happened to me.

Resting my hands on my walker, I looked outside the window. A tall, yellow building with dark red borders displayed a symbol of a burning torch and an emblem that read 'We Serve to Save'. This was the headquarters of the Kerala fire-and-rescue services. In front of the building was a mango tree with low-hanging and mouth-watering ripe fruit. I love mangoes. How I wished I could go down and pluck a few! A sense of calm enveloped me as I looked at the lush green before my eyes. As always, nature was displaying its infinite ability to restore my soul and bring me peace.

I glanced down and saw people on the sidewalks, lost in their own thoughts. I looked up to discover eagles deftly soaring to higher altitudes in the backdrop of clear blue skies and ravens happily flying against the swaying palm trees. Then, a flock of elegant white egrets with tall, slender legs descended on the mango tree, followed by another flock and then another. It seemed like some kind of a convention was underway. I also spotted a lonely grey heron sitting on a palm tree, awaiting the arrival of his friends. Suddenly, there was a rustle in the mango tree. It was a green parrot, well camouflaged with the lush green.

In the company of so many precious creatures, I would never be lonely for the next twelve weeks. These sentient beings offered profound solace and ignited the flame of gratitude. Despite being temporarily disabled, I had so many things to be grateful for—especially a pair of eyes to view this magnificent natural world and an open heart to feel my emotions. I closed my eyes and said a gratitude prayer for the treasures of the earth, then retreated into my bed, my foot elevated on the pillows. I took a deep breath and drifted back to the past few weeks.

The first fifteen days of January 2017 had brought with it many victories—and grief at the same time. On 5 January 2017, just four days before my accident, I'd had a hugely successful screening of my documentary, *Gods in Shackles.* It was held at the Kerala High Court in Cochin City. The film exposes the dark truth behind the glitz and glamour of various Indian festivals in which Asian elephants are exploited for profit behind the seductive veil of culture and religion. Among the two hundred-plus attendees at the screening of the film were judges and lawyers. They were involved with court cases that had been filed against elephant owners for their barbaric treatment of India's cultural icon. After the screening, many of these legal persons had lauded the in-depth research that had gone into my film. One of the judges had even said that it was a true eye-opener.

I returned to Trivandrum the next day, Friday, 6 January, just in time to prepare for a scheduled meeting with a top forest official. My mission was to present different ways to end Kerala's elephant slavery. I had hardly been able to sleep that night, thinking of the unlimited opportunities to collaborate with the state government and potentially create a shift in the way people perceived these incredibly intelligent and noble animals.

What if we could arrange a documentary screening for the forest officials? What if these men on the front line could use the evidence in the film to arrest the perpetrators of the elephants' abuse? What if the forest officials became inspired to host a screening for the elephant owners? They could then see for themselves the suffering being inflicted on these marvellous animals, and, thus, have a change of heart. My mind kept buzzing as I pondered all the possibilities that lay glittering before me.

On Saturday, 7 January, I had arisen at my usual hour—4:00 a.m. After undergoing my daily ritual of yoga and meditation, I took some time to reflect on the key points of the important meeting I would be having that day. Since the release of my documentary in June 2016, the forest department had been inundated with multiple petitions and e-mails from animal rights activists around the world. But ironically, one of the key decision-makers had not seen the film. Given my limited understanding of the hierarchy of the forest department, the additional secretary of forest, Mara Pandian, was not invited for the inaugural screening at Kerala's general assembly. This meeting was my chance to correct that oversight and build a trusting relationship with him.

At 3:30 p.m. sharp, my cab picked me up for my four o'clock appointment, even though the meeting's locale was a mere ten minutes from my hotel. Not surprisingly, there was heavy security all around the government building. Fortunately, a prior security clearance had been requested by Mara Pandian's secretary, and they allowed us in. My heart was pounding as I took the elevator to the third floor.

At the reception, his secretary made an internal call and escorted me into a spacious office. Behind a large mahogany desk was a man of medium stature who stood up and greeted me with a glistening smile. After everything I had heard about Mara Pandian, here I was shaking hands with him!

Wildlife photos of elephants, tigers, egrets, snakes, flowers, waterfalls, and trees were displayed wall to wall. As my eyes scanned his office, a sense of calm filled my heart. Mara Pandian explained that every weekend he would disappear into the wilderness and gather photos and videos. He then opened his laptop and proudly shared his close encounters with elephants. For a while, we had become so engrossed in his adventures that we didn't realize his secretary had arrived to take down the meeting's minutes. In a few moments, a chef walked into the office with a plate of vegetable cutlets and a tray filled with spicy cashews. I took a bite of the sizzling and spicy dish as Mara Pandian continued to share his memories with elephant herds.

Somehow, all the barriers that might have existed between us naturally melted away. We effortlessly exchanged small talk about our various hobbies and interests. He was keen enough to notice my traditional mannerisms and made it a point to bring it to my attention. With a sparkle in his eyes, he remarked, 'Despite migrating to Canada, you have retained India's cultural values.' Clearly, he had appreciated my Indian outfit and mannerisms, which included greeting him with a namaste, for instance.

Once I had established an easy rapport with him, I casually mentioned that had I known how genuinely passionate he was about wildlife, I would have contacted him much earlier. I also acknowledged that some people had kept me away from him. His piercing glance revealed a palpable relief at my words. He didn't address my comment, but instead asked, 'So what have you got for me?'

I enthusiastically opened my laptop and made a short presentation, revealing my specific plan to rescue the abused temple elephants and release them into a sanctuary. I explained how it could create ecotourism jobs for the indigenous tribal community and how, by building an Ayurvedic massage parlour at the periphery of the sanctuary grounds, we could bolster the traditional Ayurveda that had originated in Kerala. Next, I explained how the sanctuary itself could have a retreat centre for volunteers/tourists and feature a vegan restaurant. Reiterating that I was keen to work with the forest department, I also reassured him that I would do everything required of me in all future interactions with the department. At these words, Mara Pandian gave me a silent look and then flashed back a smile of approval.

Perhaps it could have been my authenticity and transparency that prompted him to ask, 'Would you like to visit the government-run Kottoor Elephant Rehabilitation Centre? We have thirteen rescued elephants there.' Obviously, I was elated by this amazing proposition to visit the centre. When I nodded an enthusiastic yes, Mara Pandian immediately phoned Shaji Kumar, the wildlife warden of Trivandrum. Together, they arranged for me to be picked up on Monday, 9 January, from my hotel. To my mind, this was a small miracle! It was indeed one of the most memorable days of my life.

Before this meeting with Mara Pandian, I had been anxious and spent hours preparing for it, only to later realize that a heart-to-heart connection mattered more than all the planning and preparation. Again, everything had flowed effortlessly. I had experienced goodness emanating between two well-intentioned individuals with a common passion. After a brief exchange of gratitude, I left Mara Pandian's office. As I was driven back to my hotel, I began to dream of being with my beloved elephants once more.

Monday couldn't come soon enough! I was ready to go by 6:00 a.m. In the hotel restaurant, as I was having breakfast, I was bubbling with excitement like a child and flashing smiles even at strangers. One of the waiters asked me what was so special about that day. I explained, giggling uncontrollably, that I was about to meet my soul animals. The story spread like wildfire in the restaurant and other waitstaff paraded in front of my table, trying to hold back their contagious grins.

After breakfast, I ran back upstairs to my room to pick up my purse as I had to buy fruits for the elephants. It was now 8:45 a.m., but my pickup had been arranged for 9:00 a.m. I still had fifteen minutes to go, and every minute felt like a long hour. Impatient, I phoned Shaji, Trivandrum's wildlife warden with whom I would spend the day. To my amazement, he had already arrived at the hotel.

At the entrance was a white government jeep with seats covered in white and green striped Turkish towels. A dark man with a pleasant smile greeted me warmly, and like a true gentleman, Shaji opened the back door of the jeep. I hopped in, thrilled to be spending an entire day with him and the elephants.

After some small talk, I began to question him about reports that the elephants in the government camps were poorly maintained. The soft-spoken Shaji politely told me, 'Madam, we're doing the best we can with what we have. The reports are sensationalized.' But of course, that response didn't satisfy this journalist. So, I hurled at him a barrage of questions that he continued to respond patiently to without ever getting agitated. In about half an hour, I had pretty much exhausted all of my questions. It was time for some silence. I can draw my own conclusions after visiting the camp.

There are approximately seven elephant camps in Kerala run by the state government, most of them maintained in appalling conditions. The Kottoor Elephant Rehabilitation Centre is the best of them all—although not the most ideal.

It has the largest space and open enclosures where some elephants are allowed to roam shackle-free during the day but chained during feeding and bathing times. Some of these elephants were abandoned by their owners, whereas most were rescued as babies from the forests. There they were reportedly abandoned by their herd after failing to save them from swamps or barbwire entanglement. Some of the older bulls had been captured, as they were deemed 'crop raiders' by the media. As of January 2017, there were four baby elephants, three subadult males, four adult females, and two senior bulls over the age of sixty-five.

A whiff of warm morning breeze carried sweet scents of wildflowers as we left behind the bustling city and drove through the countryside. We then entered a small village where hawkers were selling produce on the sidewalk. I shared with Shaji my list of fruits to purchase for the elephants. He wisely suggested that we would find fresher and cheaper fruits farther ahead.

As we drove along the dusty bumpy roads and past the gates of an amber-coloured building, loud chants of mantras echoed through large sound boxes installed at the entrance. It looked like some kind of ceremony was underway. The building was decorated with traditional orange-coloured chrysanthemum flowers and palm branches. The driver mused aloud, perhaps it was someone's wedding or some kind of religious function.

By now Shaji had spotted a large fruit stall. Here, we bought loads of pineapples, papayas, apples, watermelons, and dozens of bananas, all of which the elephants love. After stocking up the jeep with fruits, we continued our journey. Along this stretch of the road, we were surrounded by rubber and banana plantations that extended for miles. With a heavy voice and glum face, Shaji explained that once upon a time this area used to be a dense jungle, home to many different species of exotic wildlife. But humans had encroached into the zone, pushing the wildlife inside, where they are facing severe water and food scarcities. As a result, the animals are now being driven out of the forests and into the villages to find food. Elephants—considered crop raiders—are eventually captured due to public pressure. They end up in elephant camps where they are forced to live an unnatural life, tethered during their mating cycle and deprived of their primal instincts.

A few moments later, the driver informed us that we were getting close to the forest camp, and just ten minutes later, we drove past a sign that read: 'Kottoor Elephant Rehabilitation Centre'. Here, one of the friendliest people I would ever meet in Kerala greeted us with a warm smile. It was Sukesan, the forest ranger.

Straightaway, he escorted me to an open area where two baby elephants were following their handler—also their foster dad—very closely. One of them was a six-month-old male, Arjun. The other elephant was a four-month-old female named Poorna. Sukesan told me that Arjun had been found seriously injured and Poorna had fallen into a pit. After repeated efforts to save them, both elephants had been abandoned by their herds and were eventually rescued by the forest department.

These adorable, playful beings with spiky hair were mesmerizing. I asked if I could spend some time with them, and my request was instantly granted. As I entered into their space, baby Arjun came running toward me and wrapped his little trunk around my hand as though wooing me to play with him. Poorna followed him everywhere. She curiously touched my left arm with her moist, delicate trunk. As I rubbed Arjun's back, he began to doze off. The handler said this was an indication that he was feeling comforted.

Soon enough, Poorna wrapped her trunk over my head, and slid her lower jaw beneath my chin, engulfing the left side of my face. I was instantly enveloped in blissful cuteness and love. Realizing that Poorna had stolen me away, baby Arjun did everything he could to get me back and began chewing my hand. Although he was only six months old, his jaws were quite strong, and his actions certainly got my attention. Apparently, he was teething. Elephant babies are no different from human babies, who also chew and bite when they are teething. The purity and innocence of all babies—human and elephant—is out of this world.

My twenty minutes of pure ecstasy was abruptly interrupted by Sukesan. He announced that the other elephants had returned from the lake and were ready to be fed. I was directed down a trail and there they were, standing shackled in heavy chains next to one another. At the far end of the group stood two babies, a male and a female. Raja and Podichy were perhaps between one and two years old. After silently watching them from a distance, I approached them slowly, at which point, they curiously sniffed my toes. I caressed their hairy heads, which were sprinkled with mud. But they should be roaming freely with their herd, never shackled in one place. The only comforting news was, after feeding, they are released into open enclosures where they can socialize with each other. Still, this was not the same as being with their herd.

Elephants are highly social animals. Just like us, it is their primordial nature to commune in groups, graze and bathe together, and communicate in different ways, especially by touching each other frequently with their trunk. Their priority is protecting and nurturing the young, while also preparing them for the unforeseeable dangers. But these elephants would always be at the mercy of humans, depending on them to be fed and bathed, with no real chance of expressing their true nature.

It was now lunchtime, and the handlers began to feed their respective elephant large balls of rice mixed with jaggery, ragi (millet), and coconut slices. The elephants seemed hungry, for they eagerly opened their mouths to receive the proffered food. In the wild, they wander sixteen to eighteen hours a day, grazing on at least two hundred varieties of leaves, bark, roots, shrubs, grass, and other vegetation. But in captivity, they are fed every four hours—four to five times a day—barely enough to sustain their massive bodies. Furthermore, although they wandered freely in their enclosures, they had limited vegetation to graze upon.

As I was drifting into a different realm, I felt a tap on my shoulder. It was a mahout, inviting me to feed the elephants the fruits we had brought. He and the other mahouts had washed and cut the fruits into tiny bits that were elegantly placed in steel bowls. I felt rather like a mother feeding her babies as I put tiny pieces of watermelon into the little mouths of Raja and Podichy. They devoured the offering. It was then Augusthyan's turn. He was a playful seven-year-old bull elephant and when I approached him, the handlers became a bit jittery. Apparently, he was one of the mischievous ones. But to everyone's surprise, he gently opened his wide mouth and allowed me to feed him.

After rubbing his trunk and patting his back, I moved on to Unnikrishnan, who was also around seven years old. The handlers had the same concerns about Unnikrishnan, but he, too, befriended me and enjoyed being fed. Next to him was Ammu, one of the mature female elephants. She was much larger than the first four. In general, females are gentler than bull elephants, except when they feel threatened or if they have young ones in the herd. Ammu allowed me to feed her two papayas and large pieces of pineapple. I was so hypnotized by her honey-brown eyes and long lashes that I couldn't walk away from her. I said, 'Ammu, you are a good girl,' and as though acknowledging my compliment, she released a heavy breath.

I then stopped for a moment and turned around, only to discover that a massive bull elephant, the eleventh one, had just arrived. He was so tall that it would be difficult for me to reach his mouth. Perhaps after feeding the next elephant, I felt it would be best to have the handlers feed the others.

I then moved over to the last elephant I would feed that day—Rana, a four-year-old charming baby.

Suddenly my phone rang, interrupting my flashback reverie of the events of early January with the Kottoor elephants. I was jolted back to my current reality in the hotel room where I was lying on my bed. The state police chief (SPC), Mr Lokanath Behera, had promised to call me the day of my discharge from the hospital. Here he was, on the other end of the line, true to his word. He proceeded to inform me that he and his wife were planning to visit me that evening.

I made arrangements at the hotel to ensure that the couple received a warm welcome. They were escorted to my room where they found me lying helplessly; my left leg stretched and elevated on two pillows to avoid further swelling. I showed them a cell phone video, revealing how my accident had occurred. Their distress, as they watched the drama, was all too palpable. My body shuddered at the noises emanating from the phone, as it did when I replayed the scene for my brother. The video ended and silence filled the room. They looked at me in utter disbelief, trying to digest the severity of my injury. I had pre-ordered two glasses of pineapple juice. As the couple quietly sipped the juice, I reassured them that this minor setback would not prevent me from moving ahead with our plans to provide a sensitization workshop at the end of the month. The goal was to help the front-line police officers to grapple with the plight of elephants, and the existing wildlife laws, so they can enforce the laws and penalize the perpetrators. They wished me a speedy recovery and left after about twenty minutes.

The next morning, I was up bright and early after a good ten-hour sleep. I dialled my brother's room to find out if we could have breakfast together but Abhi said that Raju was still asleep. So, I ordered mine. As I awaited my breakfast, I reflected on my brother. In his youthful days, he had been quite charismatic. Having a theatrical bent, he had appeared in many plays, always captivating the audience. He also excelled in school and was a great athlete. Despite being a troublemaker, he never got into trouble; instead, I got blamed for everything he did.

After all these years, his behaviour and appearance had hardly changed. He was fifty-four years old but looked thirty-five. He had a dark moustache, a crown of naturally jet-black hair that framed his dark brown eyes, and a smile to die for. This daredevil would never hesitate to venture into uncharted territories. One of the greatest things I loved about Raju was his silly and crazy sense of humour.

He and I had a close relationship as children and as young adults. But after I migrated to Canada in 1989, we had grown apart. My visits to India had become infrequent; indeed, I only travelled 'home' every two to three years. It had been almost thirty years since my brother and I had spent quality time together. So, I was excited about hanging out with him for the next few days.

There was a knock on my door, and my breakfast had arrived. Vishnu, my favourite waiter, walked in with a tray full of Indian dishes from the breakfast buffet in the restaurant. The dosa, chutney, roti, potato masala, and hot chilli Indian pickles looked absolutely decadent. It had been a while since I'd eaten such an elaborate breakfast. In Canada, I typically had oatmeal or a peanut butter and jelly sandwich.

A few moments later, Raju and Abhi walked in, and the three of us then tucked into my breakfast, gobbling it up hungrily. Moments later their food arrived. As we ate together, we cherished our morning, joking about our restricted childhood, skipping classes, and missing school assignments. We also discussed some of our childhood nightmares, which involved a fair bit of family turmoil and domestic violence. Our father was a wounded soul who had been orphaned at a tender age. As he became older, his past demons manifested in violent ways. Despite all this, he had always wanted the very best for his children.

After breakfast, Raju, Abhi, and I played a musical game called *anthankshari*. In this game, one sings a song that begins with the last word of the previous song sung. Listening to my brother perform old classics from popular Hindi movies felt nostalgic. It had been decades since he and I had reminisced in this way. I was having so much fun with my beloved brother and nephew that my agony and suffering had eclipsed, if only temporarily.

The ten days spent with Raju and Abhi were meaningful to me during a very crucial time in my life. Those few days went by like a flash of lightning, along with the laughter, silliness, and tears . . . reigniting mutual love and respect for each other. Eventually, with a heavy heart, I had to say goodbye to Raju and his son. At the same time, I was feeling light-hearted for my brother. We had unshackled past resentments and misunderstandings and had healed on a deep level. All our preconceived judgments, resentment, hurt, and sorrows had melted away, and we were the best of friends once again.

This healing of my relationship with my brother, and my appreciation of those who cared for me after my accident would prove to be a harbinger of a shifting intrapersonal dynamic for me in the days and weeks ahead. For more than three decades, I had avoided human company, preferring instead to be surrounded by animals and the natural world. My accident was starting to change all that. Only later would I realize that it had set in motion a unique journey on which I would help to heal not only the mistreated temple elephants of India but my own isolated self as well.

Chapter 3

Synchronicities Guide Me on My Path

Four weeks after being confined to the same hotel room, my situation was becoming increasingly intolerable. Emotions were building up inside since that fateful day of 9 January at the elephant rehab centre. Drowning in depression, I began to feel victimized. Why me? What had I done to deserve such a fate? Why had I been chosen to go through this suffering? The 'should have' 'could have' and 'would have' voices were getting louder.

Unable to do anything, I had a lot of time to think. I began to ponder about the recent chain of events, wondering where my life might be leading me. But was it leading anywhere?

I have always loved my solitude. However, post-surgery, solitude had turned into loneliness and isolation; I yearned to be surrounded by people. I enjoyed their visits and would often ask them to stay longer. Sympathetic to my restricted mobility, they frequently obliged. Although I desperately tried not to wallow in self-pity, I often found myself complaining about what I deemed to be my 'house arrest'.

But my suffering paled in comparison to the plight of the captive elephants that are shackled in small areas with no chance to socialize. At least, I still had access to the outside world. I could connect with my friends and family through e-mail, social media, and the phone.

In the wild, elephants live in tightly knit families. Just

like us, their priority is protecting their young. It was hard to comprehend how babies are kidnapped from their herds by ruthless tactics. In some places, explosives are employed to scare and scatter the adult elephants, leaving the babies running helpless and confused. In other places, babies fall into pits that are intentionally dug to trap them. The mother of the calf and its relatives desperately try to rescue the little one, but after repeated efforts, they eventually abandon their young. The vulnerable babies are then captured, subjected to ruthless training, and paraded in cultural festivals under the cover of religion or exploited in zoos under the guise of education. What if someone did this to their children or grandchildren? Why and how could people be so cold-hearted and hurt these harmless and innocent animals?

The elephants used in festivals and temples have infrequent visitors and extremely limited social interaction with other elephants. They are forced to stand shackled day and night in one spot in cruelly short chains on their own urine and excrement. This not only limits their movements but also causes severe ailments. Is this not the most heinous crime against nature? It is hard to imagine what these isolated creatures are forced to endure each and every day.

My own confinement made me realize just how much my visits must have meant to the elephants—and suddenly something shifted inside of my being. Through my suffering, I began to feel the physical agonies that elephants experienced on a daily basis. I was unable to bear even the weight of the cast on my leg and the 500-gram titanium plates at my ankle. What about the elephants? The shackles on their ankles are much heavier. I began to realize just how much my current situation paralleled that of the elephants. Perhaps some invisible hand had orchestrated this series of events so that I could put myself in the elephants' shoes. Thus, I began to be more accepting of the reduced circumstances I found myself in.

Over the years, I had seen many elephants forced to

limp and parade, despite the raw bleeding wounds on their ankles. These wounds were the result of the tightly shackled chains that cut into their flesh or constant torture inflicted on their most sensitive parts. This is done using barbaric weapons such as the bull hook, or a long pole with sharp-pointed metals. These mute animals cannot express themselves or their pain in a manner that humans would understand. Their foot injury or even fracture is obvious in the way they walk. However, their plight is largely ignored, and they are typically left to suffer in silence.

My condition, on the other hand, was only temporary. My foot would heal, and I would no doubt walk again. But the elephants are doomed to a lifetime of slavery and confinement, depraved and dejected, constantly craving to be with their own kind. Short of that, any interaction or socializing would, no doubt, be welcomed by them if my visit to the forest camp was any indication. All they wanted was some semblance of compassion, someone to acknowledge their suffering, and someone to ease their burden of pain.

As I tried to come to grips with the suffering of elephants and my own physical limitations, I began to chant Lord Vishnu's prayers. I had often done this as a child. By receding into my cultural roots, I found solace. Suddenly, a gorgeous kingfisher landed atop a building across from my window. The sun's golden rays enhanced the bird's glittery shades of blue as it sat there soaking up the warmth, silently witnessing the world around him. He suddenly turned towards me, and for a few moments, it felt like he was looking at me directly. My sadness turned into joy, for, in that very instant, I felt a profound connection with life.

Lord Vishnu has a blue body, and so did this kingfisher. Pondering this, I thought that perhaps the universe was sending me a signal, reinforcing my understanding that things were happening for a reason. According to Hindu mythology, Lord Vishnu is a very popular god because he is humble and diplomatic. Perhaps there was a lesson here for me to remain humble and calm in the face of a setback to my mission. If only I could trust the universe and its wisdom, everything would work out in the end.

As I retreated to my bed, my thoughts drifted back to June 2013. At that time, my love for elephants had been rekindled through synchronicities that had unfolded. I had been on a visit to India to attend a ceremony commemorating the first anniversary of my father's death. He had died of cancer in 2012, but sadly, I was unable to attend the funeral. Thus, I returned to Mumbai in 2013 to offer solace to my family and close relatives on this first anniversary of his passing.

But I had a few days before the commemorative ceremonies. So, I made a last-minute decision to travel to Ooty and visit some family members instead of going directly to my mother's place. However, the evening before this trip to Ooty, I would be staying with my brother, Raju. He lived in a town called Bhandup on the outskirts of Mumbai. And I would fly to Ooty the very next day.

At this point in my life, I was still managing the Bermuda Environmental Alliance (BEA), an organization that I had co-founded with prominent business executives on the Islands of Bermuda. I had returned to Toronto from the island a week ahead of my trip to India so I could buy a few gifts for my family and complete some travel formalities.

A seemingly never-ending flight from Toronto to Mumbai finally landed at the Chhatrapati Shivaji International Airport, which was bustling even at 12:30 a.m. It felt like a furnace, as I stepped into the hot and humid airport. Outside, my brother, Raju, was waiting for me amid a sea of people, but certainly not as chirpy as usual. With a dejected look on his face, he hugged me and then proceeded to take my luggage. I quietly followed him to his car.

During the drive, my brother shared what had been his last few interactions with our ailing father the year before. Raju's voice trembled as he explained that just two days after he had returned to Qatar, he received the devastating news that our father had passed away. Raju seemed just as guilty as I did for not being at our father's side when he died.

Silence fell over us as my brother drove to his flat in Mumbai. Even though it was very early in the morning and the roads weren't as busy as they would be during peak hours, we encountered reckless drivers and people crossing the highway in a random fashion. Enormous billboards that promoted Bollywood movies, the hallmark of Mumbai, were decorated with gaudy flashing lights. By the time we got to my brother's place, it was almost 2:00 a.m. After an exhausting journey, I barely had the energy to greet my sister-in-law. I was ready to crash, yet had to be careful not to sleep in, as I had a flight to catch in a few short hours.

Despite the sombre reasons for my visit to India, I was thrilled to be returning to the gorgeous hill station of Ooty after having been away for four years. My uncle, Mohan, came to pick me up at the Coimbatore airport. From there, we drove about eighty-five kilometres to Ooty. As our van began ascending the hairpin bends, I was awestruck by the breathtaking beauty of the mountain ranges covered in mist, and my body began to thaw. Further up the hills, waterfalls gushed abundantly. A cool, crisp breeze caressed my face, and I inhaled the soothing scent of eucalyptus. Monkeys hopped from tree to tree on the canopies that covered the road.

I can't explain why, but I suddenly thought of my old friend, Raj, a photographer and wildlife conservationist. I hadn't connected with him for decades and asked my uncle whether or not he had been in touch with him. Without a moment's hesitation, he phoned Raj. As fate would have it, just that day, Raj had returned from a distant field trip related to his work in tiger conservation. As things turned out, he would drive through the pouring rain that same evening to visit me.

The minute I saw him, precious memories of past times spent together flashed before my eyes—delicious meals, laughter, and the good times we had shared. Raj was a long-time family friend. We had met when I was in my early twenties. After I moved to Canada, however, I'd neglected to stay in touch with him. After all these years, we picked up where we had left off, without any guilt trips or complaints.

During our brief visit, I shared with Raj that I was a nature and wildlife journalist and was keenly interested in protecting wildlife. I also subtly mentioned how much I loved elephants and gently asked if he might have some time to take me to a nearby elephant forest range—the Mudumalai National Park and Wildlife Sanctuary—the next day. He smiled and casually said that he would take me to an even better place, a dense jungle where elephants thrived.

He was talking about Wayanad in the southern Indian state of Kerala, a key global habitat for Asian elephants. Coincidentally, I had been born not too far from Wayanad, in a little village named Alathur in the neighbouring Palakkad district. My family had left Kerala when I was three years old. However, even after so many years, I still vividly remembered how my grandparents used to take me to a temple where there was a stunning bull elephant with whom I played. His handlers had never been concerned about leaving me alone with this gentle soul. In the many decades that followed, I had frequently visited various parts of my home state of Kerala but hadn't returned to the village of my birth.

Even though I was staying in Ooty for only five days, my uncle knew that I was torn between spending time with elephants and the family. He, thus, gave me his blessing to spend the next afternoon with Raj. That night I kept tossing and turning in bed, thinking about Kerala and my connection to it . . . the elephants!

As usual, I was up before 5:00 a.m. and after my morning rituals and some coffee, I was ready by 7:00 a.m. Raj phoned and said he was running late as the roads were jammed with traffic caused by the torrential rains and flooding the previous night. I ran up to the terrace, anxiously awaiting his arrival. A little while later, I spotted his white Subaru, and in a matter of seconds, we were off on an adventurous day.

The smell of the earth carried by the cool morning breeze seeped through my nostrils, awakening every cell of my being. The monstrous dark clouds floated in the skies. Thunder rolled over us, threatening another bout of a heavy downpour. Life was bustling on the narrow streets. Men were herding cattle, women carried mud pots filled with water on their heads and hips, and children in uniform were on their way to school. Along the roadside, wild hens were pecking away the morning worms from the red earth, oblivious to the chaos on the main street.

Suddenly, my attention was diverted by a mama monkey with a baby in her arms. In that instant, my heart skipped a beat as I watched her cutting across the traffic to grab some bananas that a driver was offering. Although well-intentioned, Raj lamented that such behaviour toward the animals has resulted in hundreds of senseless monkey deaths. He said that these irresponsible acts have fostered codependency between humans and non-humans. Fortunately, this particular monkey managed a narrow escape.

By now, it was almost 9:30 a.m. After driving down the slushy and winding roads of Ooty, my stomach was feeling queasy. So, we pulled over for some coffee and breakfast in a little town called Mettupalayam where a popular restaurant was packed with people. There, I waited patiently for my favourite filter coffee and dosa (pancakes made of rice and lentils). This was served with sambar (gravy made of lentils, vegetables, and spices) and chutney, which I savoured.

Soon after breakfast, Raj had to stop at his wildlife conservation bureau to check in on the tiger project he was working on. As we drove through a narrow stretch of bumpy road to his office, it was hard to ignore the open sewers that lined the street. Large vehicles and scooters from the opposite direction were trying to push through the tapered lane. At one point, my body and mind couldn't handle the chaos—the sensory overload was just too much.

My head began to spin, and I had to close my eyes and take a deep breath.

The reality is, this is how millions of people live in India day in and day out. In fact, I, too, was part of such an existence at least for the first twenty years of my life. No doubt, the place had become overcrowded and things had changed quite a bit. But it became clear to me that after moving to Canada, I had become disconnected from the realities facing people in India.

Somehow, we managed to reach Raj's office in one piece. As he entered the two-storey building, I went up the stairs to an open terrace. The lush scenery was breathtaking! The sun was finally breaking through the clouds. It cast its warm rays on paddy fields that stretched for miles on end. The smell of warm earth, emanating from the mixture of sun and rain, calmed my senses. The exotic melodies of India's endemic mynahs filled the air. As a young girl, my grandma had taught me that these clever birds can mimic human speech, and so I tried to talk to them. But they flew away, probably laughing their hearts out at this stranger's language. In the middle of the field, I spotted a white heron elegantly standing on one leg in a yoga posture. Everything felt serene, beautiful, and in perfect harmony. I couldn't help but take many scenic photos.

An hour later, Raj reappeared after briefing his team on an upcoming wildlife initiative that they were undertaking together. And now he was free to travel to Wayanad with me. As Raj backed up his car, there was a big splash! It slipped into the ditch that we had parked beside. I shook my head in utter disbelief—there was always something! And yet it was not too long before the locals gathered around the vehicle. The whole community was coming together to help us. After twenty minutes of struggling to lift the car from the ditch, it again slipped from their collective grasp. At this point, they all looked at each other with utter despair, as though they had given up. But in a matter of seconds, they lifted the car from the ditch and dropped it with a heavy thud back on the road.

After an exhausting ordeal, we were finally on our way to Wayanad. However, within just thirty minutes of the drive, Raj's phone rang. He pulled over to answer it. The voice at the other end of the line was serious. It turned out to be a distress call from the Wayanad wildlife warden. Apparently, the forest officials were desperately trying to rescue an elephant that had slipped and fallen into a trench. It had been built along the fringes of the village to deter marauding animals.

This mighty bull elephant was known to frequent the village and graze on the villagers' crops. He had tried to enter the village the previous day after dusk. However, the red mud had become too slippery after the monsoon showers, and he had fallen into the ditch. Raj swung into action and phoned his wildlife bureau to order up a GPS collar to install around the elephant's neck. This was necessary to track the elephant's movements after his rescue and ensure that he wouldn't cross over the trench again and get into yet another accident. So, we had to return to Raj's office to get the collar and then drive to where the tusker was struggling to get out of the ditch.

Ironically, the area of the jungle where the elephant had fallen was the same place that Raj and I had intended to visit that day. The sense of surrealism that I was experiencing was heightened by the fact that the car Raj was driving had suffered the same fate—that of getting stuck in a ditch—as the elephant.

At the time, these happenstances seemed like a meaningless coincidence. Yet I have since learned that there are no coincidences in the universe. It would, in time, become evident that there was a reason for the events that unfolded and the way they did. I would be reminded to trust in the universe and its innate wisdom in my fate's unfolding.

Raj turned the car around and drove back through the crowded streets, dodging the hawkers and autorickshaws—those pesky three-wheelers painted in yellow and black that buzz like motorized bees. We were now back on the same narrow lane we had been on before. Raj was driving extra cautiously so that his Subaru wouldn't slip into the ditch again. Fortunately, we did not have to park the car this time. Instead, we pulled over and a member of Raj's conservation team hopped into the vehicle with the GPS collar.

We drove silently along the slushy road on a seemingly endless drive, my gaze resting on the lush green forest patches interspersed with vast croplands that surrounded us. This was offset by the backdrop of the mountainous horizon. It began to drizzle again, as the raindrops carried by a brisk wind stung my face like the prickly pines.

Throughout the drive, we had to stop frequently and ask for directions as there were no traffic signs in sight. Still, we lost our way a few times before reaching our destination. This was the little village where, at least, a thousand people had lined up on the sides of a muddy trail to witness the drama unfold. Crowd control is a huge issue in India as the human population is swelling at an unprecedented rate. Police were conducting spot checks at the entrance to a pathway that led to the ditch with the elephant in it. This would help to control the crowd. But they allowed us in after Raj showed them the collar and explained to them that we were part of the rescue team.

As we drove cautiously through the crowd, a potent stench of manure filled the air. Everything around us seemed like a mirage—the rain, the slush, the crowd, the loud chatter, and the powerful odour filling the air. Amid all the chaos, a helpless animal was lying at the mercy of humans who would not hesitate to kill him if they felt threatened in any way.

We gathered our heavy equipment—the GPS collar, battery, and tools—and made our way to the trench. I followed Raj gingerly through slippery puddles, brushing past the noisy crowd. We were now confronted by a human wall of wardens and officers who had cordoned off the area with yellow 'caution' tape. I hastily walked past one of the wardens and peered into the ditch. And there he was! The magnificent bull elephant whose coveted tusks were covered in slush and red earth. He was lying in the trench in a precarious position. His front feet were flung up in the air, his rear legs bent, and his back was squeezed between the narrow walls of the trench. His eyes were wide open and alert, as his trunk wiggled restlessly.

The first step in freeing the bull was to widen the trench. To do just that, an excavator moved into the zone, the loud sounds of its obnoxious engine reverberating throughout the forest. The poor elephant was visibly terrified, displaying symptoms of trauma—his trunk moving rapidly and his feet kicking the air. The only way to relax him was with tranquillizers. A veterinarian dedicated to rescuing and rehabilitating elephants shot a long, thick dart into one of the tusker's hind thighs. And so began a delicate rescue operation as the giant drifted off into slumber.

Now we were in for some nail-biting moments. The operation resumed as the excavator began digging out the trench wall to widen it. The scariest moment for me was watching the vicious claws of the equipment land right next to the elephant. At one point, I yelled, 'Watch out for his ears!' The warden gave me a cold look. I sheepishly apologized and tried to remain calm.

As I looked around, I realized that I was the only woman standing so dangerously close to the trench. I struck up a conversation with the wildlife warden, who introduced me to one of the key veterinarians. I shook the vet's hand and instantly heard whispers in the background. I was in a primitive village where it was highly unusual to see a woman dressed in jeans and a sweater (Western garb), to say nothing of the fact that I was shaking hands with a man. Perhaps after having lived in Canada for decades, my Western persona had taken over. Indeed, I had no inhibitions speaking to strangers.

I surveyed the crowd and realized that most of the men wore white sarongs that were folded up to their knees. My father also wore a white sarong as he sat before our wooden mini temple at home to do *pranayama* (single nostril breathing) and worship with his eyes closed. He offered fruits to Lord Ganesh, a popular Hindu god with an elephant face, considered to be the remover of obstacles. I closed my eyes and took a deep breath, seeing my father, through my mind's eye, lighting incense sticks and camphor.

Growing up in a Brahmin family, I had watched my parents perform rituals daily before they left for work. They taught me to pray every morning and evening and to be pious before elders. It was sobering to realize that, over the decades, I had changed so much. I felt like a virtual stranger in my own homeland.

Suddenly, I heard men running toward the trench, carrying medical equipment and the collar. It had been more than twenty minutes since the elephant had been tranquillized. They quickly determined that he was fully unconscious now. Everything slipped into slow motion as I witnessed the veterinarians run down the trench and Raj's coordinator climb onto the elephant's belly to install the GPS collar around his neck. First, they had to slide the collar under the elephant's neck, pull it from the other side, and then buckle it around. But the animal was covered in slush, making it slippery and difficult to put it on.

As I peered into the trench, the elephant was still motionless. At least three vets were monitoring his condition. The tension was mounting as members of the rescue team had to install the collar before the elephant regained consciousness. Meanwhile, in the background, the policemen were trying to manage a crowd of villagers that was swelling by the minute.

After fifteen minutes of intense operation, they discovered that the collar was too long. So, they had to unbuckle it, pull it out, cut it short, and then go through the same process all over again! The men standing next to me shook their heads in utter disbelief and exasperation. Added to this, when the collar was put on for the second time, the elephant became fidgety. The sedative was losing its effect. The men panicked but somehow managed to install the collar and scrambled back out of the pit.

Soon the chief vet injected a shot of an energy booster into a vein in the elephant's left ear and then painstakingly climbed out of the slippery trench. I suddenly noticed a sedation dart on the elephant's body that they had forgotten to remove. I panicked and instantly reported my observation to the wildlife warden, who was standing right next to me. He instructed the vet to remove the dart immediately, averting a serious disaster. What if the bull had been unable to balance himself after regaining consciousness and fell on that side? The dart could have penetrated his body and wounded him seriously. Placing my hand on my chest, I breathed a sigh of relief. I glanced at the warden and he nodded his head, as though acknowledging my gesture of gratitude.

By now the villagers were getting restless about the elephant's revival. They gathered around the animal and began to make sounds so that he would be energized to muster up the strength to rise to his feet. It felt surreal witnessing such unconditional love being bestowed upon an animal that they also loathed when it grazed their crops. During those moments, the human-wildlife conflict melted away. I remember thinking in Africa, where poaching is rampant, the elephant could have been instantly killed and his ivory tusks extracted for the illicit ivory trade.

Nearly twenty hours had gone by since the elephant had fallen into the trench. He hadn't even had a sip of water since the previous night. Elephants generally need to graze and browse for at least sixteen to eighteen hours per day and drink 200 to 250 litres of water daily. The poor animal must have become weak, both physically and emotionally. Perhaps he was too weak to rise up and then stand on his feet.

I looked to the heavens for some comfort and caught a glimpse of monstrous clouds sweeping across the ominous skies, thunder hovering over our heads. Another heavy downpour at any time would make the rescue operation very treacherous. The life of this magnificent animal was now in the hands of Mother Nature.

More tense moments ensued as everyone held their breath, waiting for the elephant to wake up. The booster shot was supposed to take effect in a matter of fifteen minutes. Yet twenty minutes went by and then thirty. By now the elephant was fully conscious but unable to move. After the longest forty-five minutes of my life, he moved his trunk slowly. Then his legs began to twitch. Finally, gathering all his strength, the embattled elephant rose to his feet. He turned in the direction from whence he had come and took only three steps before he lost his footing and landed on the loose mud with a heavy bang. There, he lay helplessly, his massive belly rising and falling with each laboured breath.

The wardens darted another tranquillizer into the mammoth so that they could enter the trench to administer a second booster shot. After ensuring that the elephant was sedated, two vets walked down the slope. In the background, the demolition crew began to remove the loose mud. They also tied a rope from a tall winch to the elephant's collar to support him when he tried to stand up the next time. It took at least half an hour for the vulnerable pachyderm to finally rise to his feet. After taking a few small steps, he collapsed again. Almost twenty-four hours had gone by since the accident. The sun was dimming, and so was my hope.

As I glanced at the veterinarian, he was shaking his head hopelessly. He paced back and forth with a deflated look on his face. The warden said that the elephant might have sustained serious back and neck injuries that could turn fatal. At that moment, Raj rushed towards me. With a worrisome look, he told me, 'We must leave, as you won't be able to handle what might come next. Besides, it's getting dark, and I want you to see some elephants in the wild. That is, after all, what we have come here to do.' He was gentle and coaxing. Knowing that he had my best interests at heart, I reluctantly acquiesced. As we walked towards the car, I kept looking back at the brave vet standing beside the elephant, unwilling to give up.

Raj and I drove off with heavy hearts, leaving the commotion behind. We passed through a wildlife corridor and drove into the Mudumalai conservation zone. Along the way, we saw seven elephants happily grazing with their herd on the side of the highway. Hundreds of spotted deer huddled together and glanced at us curiously. But these magnificent sightings did little to console my restless heart. The fate of the fallen elephant continued to haunt me. I made several phone calls to the man who had collared the elephant in an attempt to determine the animal's fate. Alas, they went unanswered!

Finally, at around 7:30 p.m., as we were driving up the hairpin-bend roads near Ooty, Raj's phone rang. I answered it. A familiar voice on the other end of the line said, 'Just moments ago, the elephant returned to the jungle.'

I couldn't have asked for a happier ending to one of the most adventurous journeys of my life! As Raj and I were driving, I kept thinking about the love-hate relationship between humans and the wildlife. The villagers who wanted the elephant rescued had been indirectly responsible for his fall. After all, they had dug up the trench to prevent the wildlife from entering into their village. Yet, they were capable of showing great love to these creatures and, apparently, wanted the best for them. They were also capable of treating them quite cruelly when they thought the situation demanded.

This love-hate relationship was not unlike the dynamic that my brother, Raju, and I had been subjected to by our father. Again, the parallels between the elephants' lives and my own were not lost on me. To my young mind, it felt like these coincidences could not be mere random, isolated happenings.

I was scheduled to return to Mumbai from Ooty the next day. Even though I planned to stay in touch with Raj, I was worried that it might prove more difficult for me to keep up with the elephant's fate.

During my flight to Mumbai, I couldn't stop thinking about these mysteries and coincidences that this visit to India had offered me. I was ostensibly there for my father's death anniversary. Yet why had I spontaneously decided to visit Ooty? Looking back, I now realize that it was also not a coincidence that Raj had just returned from his travels the same day I was in Ooty. And why had he offered to take me to Wayanad of all the places? This was a place that I had a deep personal connection with. It is also worth noting that the elephant had fallen into a ditch in the same jungle we had planned to visit. Then there was this parallel happenstance: Raj's car falling into a ditch and the elephant falling into a trench. And why had all those men in white sarongs conjured up memories of my father?

As I continued to reflect on how these happenings had occurred, I came to believe that they were perfectly orchestrated. Later, I would discover and embrace the great Swiss psychologist Carl Jung's theory of synchronicity. Jung posits that seemingly random events, which apparently have no dynamic of cause and effect, can be connected or related by some form of a meaningful association.

Given that the events had transpired on the occasion of my father's death anniversary, it seemed as if my father's spirit had been guiding me to and through their unfolding. Perhaps after moving to Canada, I had forgotten my innate bond with nature in general and elephants in particular. By directing me to the fallen elephant in Wayanad, maybe, my father was trying to reconnect me with my cultural roots and with my soul animal.

This mysterious encounter with the fallen elephant left an indelible impression in my heart. It was the first of many other inexplicable events that would start to crystallize my life's direction. I still wasn't so sure what my true calling was. But in this moment, I sensed that a footpath was opening up before me, which would become increasingly visible in the coming days and weeks.

Chapter 4

A Sense of Mission Emerges

As the months of 2013 went by, my encounter with the fallen elephant of Kerala continued to haunt me. I was now back in Toronto but stayed in touch with Raj. It was deeply gratifying to learn from him that the rescued elephant was never seen in the village again. But then suddenly, the elephant's movements became untraceable and the worst was feared. However, a few days later, some villagers spied him near the same trench that he had fallen into. Although this time, he would not dare to cross it. Eventually, his collar was retrieved; somehow, the wily old elephant had managed to discard it!

As I mulled over all I had witnessed in India, I became determined to learn everything I could about Asian elephants, both captive and wild. Of course, growing up in that country, I had absorbed a lot about them. I knew the basics. For instance, many zoo elephants are Asian elephants and that the Asian elephant is much smaller than its African counterpart. Most Asian male elephants have tusks, whereas the females do not. But there is always so much more to explore and learn.

In 2010, it was determined that the world's elephants are broken down into three distinct species—the Asian elephant, the African savannah elephant, and the African forest elephant. The Great Elephant Census (2007–2014), funded by Paul Allen, the co-founder of Microsoft, revealed that only around 350,000 African elephants remain in the wild, down from several million at the turn of the twentieth century.

Regarding the Asian elephant, there are approximately forty thousand left, sixty per cent of which are found in India. Additionally, 2,500 Indian elephants are in captivity, most of them found in the Indian states of Assam, Kerala, Tamil Nadu, Rajasthan and Gujarat. At the same time, the human population has been increasing dramatically in India. As of June 2020, there were 1.38 billion people, according to Worldometer, and the population is projected to swell to 1.4 billion by 2027. India is well on its way to becoming the world's most densely populated country on the planet.

In order to sustain the burgeoning human population, state governments are engaged in unprecedented development projects, agricultural plantations, dams and canals for irrigation, roads, and railway lines. The problem is, they cut through core elephant habitats, fragmenting the vast forests, disrupting their migratory routes, and pushing them out of the forestland.

At the same time, the significant growth in the human population is driving disadvantaged people towards the forest fringes. These people not only encroach into the wild habitat to gather natural resources but also to cultivate crops. Given that the fodder is shrinking inside the forests, naturally, elephants are drawn to the crop fields near the forest fringes where they can find delicious and healthy food. This, in turn, has intensified human-elephant conflict. In extreme cases of conflict, elephants are caught and taken to the so-called elephant training centres where they languish, lost to the wildlife population.

It is noteworthy that the relationship between humans and elephants evolved differently in Asia compared to Africa. In fact, throughout Asia, elephants have been revered since time immemorial. Their intelligence was venerated so much that elephants were even used to select kings. Centuries ago, elephants were used in warfare when people fought with swords and shields. These animals were also exploited by the timber industry to haul massive logs from the forests as mechanization was not advanced enough.

However, in this modern era, sophisticated machines and weapons of destruction have replaced the elephants, rendering them indolent. So, what could be done with all these idle captive elephants? A brilliant plan was devised! Religious institutions entered the scene and began to utilize them in cultural and religious festivals.

Historically, elephants have been associated with royal families, where these animals were treated with extreme care and compassion. And, although kings and emperors were using elephants for temple rituals centuries ago, the practice of exploiting hundreds of elephants is fairly new. Over the past five decades, people have been commercializing these animals, blinded by their insatiable drive for material wealth and status quo. This happens in temples behind the veil of religion, in zoos in the name of education, and in circuses for the sake of entertainment where they are forced to perform unnatural tricks. As a tourist attraction, they are forced to give rides on their tender spine, tortured to paint, and participate in races, while being deprived of food, water, and shelter.

The physical and psychological traumas that Asian elephants endure in these settings are beyond anyone's comprehension. First, they are ripped apart from their tight-knit families and stolen from the wild, then they are brutalized to submission, and then forced to perform unnatural tasks so that people can make money. By this, they are 'domesticated' in a way that the elephants of Africa have never been.

In the ensuing months, Raj continued to send me the latest studies that shed light on the unimaginable suffering of 'temple elephants'. One such study by a man named Surendra Varma not only supplemented my limited knowledge about temple elephants but also resonated deeply. This study pertained to privately owned elephants in Kerala that were used in cultural festivals. It was a rare analysis that depicted gruesome images and heartbreaking tales of captive elephants being despicably treated.

Although elephants are used in temples in a few states in India, the sheer number of elephants employed in Kerala is staggering. Not surprisingly, the festivals in Kerala had been increasing steadily since the 1980s. Around this time, most of the state's employable men migrated to the Middle East to work on oil and gas rigs there. They made loads of money and when they returned home, they purchased elephants as a status symbol.

There are, at least, three thousand temple festivals across the state every year. Thrissur Pooram is, by far, the biggest. In it, bull elephants are decorated and paraded on melting tar roads beneath the scorching sun, forced to hold their heads up for hours on end, displaying their majestic tusks. Most of these animals have been illegally bought and transferred from the north Indian states of Bihar and Assam.

In addition to receiving information from Raj, I also conducted my own online research. In 2016, Kerala housed more than six hundred of the approximately three thousand captive elephants of India, making up for around eighteen per cent of the entire captive elephant population of the country. But by the end of 2019, that number had declined to less than five hundred.

According to the media, a total of more than seventy captive elephants died under 'unnatural conditions and at a young age' in the Indian states of Kerala, Tamil Nadu, and Rajasthan between 2015 and 2017. In Kerala alone, a record thirty-four captive elephants had died in the year 2018 (almost three elephants per month), most of these deaths caused by torture, neglect, overwork, or faulty management practices.

A deeper delve into the issues revealed that this abusive behaviour had been eclipsed by glamorous displays of majestic elephants as they dutifully performed the temple ceremonies. Meanwhile, elephant owners make as much as 7,00,000 rupees (approximately $US10,000 or £7,530) per day in exchange for 'their' elephant's participation in a temple ceremony. This is appalling as elephants are considered to be the embodiment of Lord Ganesh, who is revered around the world.

Ironically, despite the elevation of the Asian elephant to India's Schedule 1 and heritage animal status, afforded in 2010, their torture and neglect continue unabated. The Wildlife Protection Act of 1972 established Schedule 1 as the official list of wildlife species at risk. It classifies those species as being extirpated, endangered, threatened, or of special concern. Once listed as Schedule 1, the measures to protect and recover a listed wildlife species are implemented. Clearly, as Schedule 1 animals, elephants of India were not being afforded the protection that the law demanded and what they deserved.

Greed and selfishness fuel such despicable behaviour and malign the Hindu tenets of compassion, kindness, and love for India's cultural icon. It is outrageous, to say the least. But, apparently, the public was unaware of the dark truth shrouded by the glamorous displays of ill-treated and abused elephants, while some people were turning a blind eye to their suffering.

The more I learned, the more I wanted to learn, and on top of this, I was impelled to act. Thus, I embarked on a two-week journey to Kerala in December of 2013. I had a new context from which I could view the captive elephant situation in India, and I intended to take things to the next level by way of my own personal observations on the ground.

In this, Raj helped to arrange some meetings for me with some of India's most renowned elephant experts. My instinct was that I should also employ my film-making skills in some way. Thus, I planned to bring my video camera and all its attendant equipment in order to capture on film material. Raj had also asked some of his younger colleagues to assist us in these research excursions.

At this time in my life, my work in Bermuda involved

a lot of travel. I'd work on the island for three weeks, then return to Toronto for a month or so where I continued to work remotely. Given that there were no direct flights from Bermuda to India, I had to return to Toronto—from Bermuda—to fly to India.

I landed in Bangalore, the Silicon Valley of India, for a day of meetings before leaving for Kerala. That night, I checked into a prearranged guest house and kept thinking about what lay before me. I was elated and nervous at the same time. I took a deep breath, listening to some relaxing music on my iPod, and drifted off to sleep.

I was up by 4:30 a.m. and opened the windows, allowing some cool air into my room. I could hear the birds chirping. However, the dark skies and empty streets reminded me that the night was still lingering. I did my morning routine, listening to a meditation soundtrack. But I was unable to focus. Two voices in my mind were restlessly battling. One voice told me that I wasn't prepared for this journey and the other one commanded me to trust and go with the flow. Breathing deeply, I silently witnessed my internal turmoil. I was reminded of Mahatma Gandhi's words of wisdom: 'When doubts haunt me, when disappointments stare me in the face, and I see not one ray of hope on the horizon, I turn to Bhagavad Gita and find a verse to comfort me; and I immediately begin to smile in the midst of overwhelming sorrow.'

As I had mentioned previously, growing up in a Brahmin family, we had chanted Sanskrit prayers routinely. We had also frequently read the Holy Bhagavad Gita, the ancient and revered Hindu scripture that is part of the epic story of Mahabharata. But after moving to Canada, all of this had fallen by the wayside. When I had visited my ailing father in 2008, he surprised me with a precious gift—my own copy of the Bhagavad Gita. This helped me rekindle my bond with my cultural roots. Ever since then, I've been soaking up its ancient wisdom.

Fortunately, I had carried this comforting companion

with me. I stopped my meditation and picked up the holy book, turning to Chapter 2, verse 31—the wise words of which always empowered me during times of inner turbulence. It read, 'Looking at thine own duty though ought not to waver, for there's nothing higher than a righteous war.' The message became clear—no turning back for me!

Till this point in my life, I still hadn't determined what my duty or dharma (life's purpose) was. But I was becoming acutely aware of the horrible plight of captive elephants in India, and I felt increasingly drawn to help them. It was beginning to appear that I was meant to bring awareness about their suffering to the world. In doing so, I hoped that their deplorable conditions would improve, and their agony would be mitigated.

While I might have suspected that this was to become my life's work, there were so many unknowns before me that made me tentative and afraid. I knew deep inside that my role was to surrender my ego in all of this. I needed to learn to continue to trust in a higher power. I also needed to trust that the direction in which I was being led would become clearer to me as time went on. This was a difficult thing to do because I had always prided myself on having a strong ego and will. These were the tools I had employed on my path to success as a broadcast journalist.

I brought my wandering thoughts back to the comforting verse of the Holy Gita and focused on surrendering to a higher power by taking another deep breath, allowing my clouded mind to settle down. The next thing I knew, the clock chimed—it was 5:30 a.m.—and as I peered out the window, there was now as much clarity outside as there was inside my mind. Doubts had drifted away with the darkness, and I enthusiastically greeted the dawn.

Shortly thereafter, I was ready for my first meeting. My room phone rang at around 8:00 a.m. informing me that Surendra Varma had arrived for our breakfast appointment. He was one of the few elephant researchers in India who had published several articles about the plight of Asian elephants, including Kerala's captive elephants. I briskly walked down the stairs of my guest house, eager to meet this man for the first time.

Varma was over six feet tall with distinct almond-shaped eyes and a salt-and-pepper beard. He seemed humble and simple, a man of few words, and a keen observer. By his poise and calm appearance, one could conclude that he was the analytic kind. He had over three decades of on-the-ground experience with elephants, and I was thrilled by the opportunity to pick his brilliant brain.

We made our way to the restaurant on the same property and settled in for a hot breakfast. However, before I could begin our discourse, I was distracted by the familiar smell of south Indian flavours that filled the air. I felt nostalgic thinking of my beloved grandmother who used to make the most delicious *upma* (a kind of spicy pudding made of semolina and vegetables, as well as chutney). The waiter served the authentic dish and filter coffee that are popular in southern India.

As I inhaled the delicious scents and then demolished the breakfast, Varma began to share his adventures and deep love of elephants with me. He made me realize how disconnected I had become from India's natural heritage. I knew then and there that this trip to India, aiming to educate myself about its native elephants, would, in fact, be an intense learning journey on many other levels, too.

I suddenly found myself in the middle of an impassioned debate about the plight of India's captive elephants. At the end of it, Varma and I were able to agree that humankind's ignorance and wilful blindness were root causes of the negligence and brutality that were routinely inflicted on these sentient beings. From what I had read, not much was being done to curb the inhumane activities. The legislation was in place but there were several loopholes in the system. This made it easy for the abusers to dodge their way around the rules.

It was deeply disturbing to learn that some of Kerala's politicians were not only elephant owners but also sat on the board of elephant owners' organizations. This is a direct conflict of interest, for in these capacities, the opportunity clearly existed for them to promote their own personal interests before the elephants' welfare. This only perpetuated the unnecessary suffering of these innocent animals.

Varma explained that much of his research being done on India's elephants was being used for policymaking. Yet, despite this assertion, through our conversation, one thing became increasingly clear to me—the well-substantiated research that Varma and his colleagues were generating on a regular basis was not being disseminated effectively to those individuals and institutions who could benefit from it the most. Another point of concern was that mandatory education and training for key stakeholders who managed elephants did not exist. I also sensed a wide communication rift between the scientific community and the public.

Varma and I became so engrossed in our fiery discussion that I didn't realize it was already 9:30 a.m., and that I needed to get to my next appointment. Varma had arranged a meeting for me with a world-renowned elephant scientist who was also a professor at the prestigious Indian Institute of Science (IIS), a nearby university. Finishing up breakfast, Varma and I left the café together, and I rushed to my room to gather my camera gear. A few moments later, we were on our way to the institute where I was hoping to conduct an on-camera interview with Dr Raman Sukumar, the head of the ecological sciences department at the university.

Given that it was rush hour, the traffic was chaotic. And as Murphy's Law would have it, we were delayed by the time we reached the university gates. Here, a steady stream of students was crossing the road at a leisurely pace, and we had to slow down to allow them to do so. Finally, we reached the parking lot and then made our way to the university's front entrance.

It was an open area where ancient banyan trees and rose shrubs thrived in perpetual sunlight. The fresh smell of red earth and the sweet scent of exotic flowers greeted us as we made our way into the main building. Students were hanging out in the lobby and professors walked to and fro—the whole place was a beehive of activity. Varma and I began climbing up a long flight of stairs, huffing and puffing as I lugged the camera equipment. When we reached the office, Dr Sukumar's secretary greeted us and informed that he was on his way. We had made it in time, after all.

We were ushered into Dr Sukumar's office, where an unusual portrait on the wall caught my attention. It was an acrylic painting of a mesmerizing honey-brown eye of an elephant glancing directly into mine through prominent dark eyelashes. I don't remember how long I stood there, captivated by the masterpiece, oblivious to Varma's presence. The next thing I knew, Dr Sukumar had walked in. The tall, lean professor, with his distinct crown of salt and pepper hair and silver-grey moustache, seemed well-grounded. Smiling at my fascination with the painting, he explained that his daughter had created it and gifted it to him for his birthday. Because I had read so much about this man and we had exchanged numerous e-mails prior to my trip, he and I enjoyed a sense of familiarity. But I was also nervous.

My goal was to interview Dr Sukumar, instead, he drilled me with questions, perhaps, trying to determine the sincerity of my mission. Dubious about the news media, he was clearly sceptical about meeting this journalist. But, soon, he relaxed. He then glanced at my camera gear and asked when and where I wanted to interview him. I promptly informed him that I had found the perfect location during our drive through the campus, and that I was hoping to interview him the same morning as previously planned. But to my great disappointment, he said that he had just returned from an exhausting trip and asked me to contact him when I returned to Bangalore during my second scheduled visit of the year. This was due to take place in the latter part of December. After a few more casual exchanges, we scheduled a tentative date for an interview and parted ways. On our way out, my camera gear felt a lot heavier than it had that morning. Perhaps the weight of disappointment had added to it.

As Varma and I walked down the stairs and made our way through the canopies to a campus cafeteria, a whiff of warm, gentle breeze carrying the aroma of wildflowers caressed my skin. Sweet melodies of rare birds echoed through the open space. It was nearing midday and the sun's intense rays danced through the canopies. With a cup of coffee, we sat on a bench under a banyan tree as Varma shared some of the intricate details of his own research on captive elephants. This study had involved spending several days with various elephant owners, handlers, and temple officials across Kerala, interviewing them, listening to their grievances, and cultivating relationships with them. As I listened keenly, it became clear that he could produce his groundbreaking report by listening intently to the elephant owners' and handlers' sides of the story and, eventually, winning their trust.

As a journalist, I knew that relationship-building was important. But in keeping up with the twenty-hour news cycle, I'd seldom had this luxury. Usually, I was in and out of the scene of a news story in a matter of minutes, leaving little time to nurture a relationship or win the trust of those I was interviewing. Of necessity, my focus was on the crux of the story and any unique angle I might be able to bring to bear in order to beat out my competition.

Given my failure to interview Dr Sukumar, it suddenly hit home that it might well prove to be very challenging to gather information on this trip, especially given that my entire stay in India was planned to last only thirteen short days. But still, Varma and I rolled out a tentative plan outlining what could be accomplished within the limited time.

It was now past noon and my flight to Kerala was scheduled to depart at 5:00 p.m. The traffic was light, assuring that we would get there on time. But I was shocked by the number of people riding bikes and mopeds without helmets. It was particularly alarming to see a woman without a helmet riding on the back seat of a scooter, carrying an infant, around three months old. The baby was fast asleep on her shoulders, oblivious to the obnoxious horns. As my cab overtook them, I kept turning back, worried for the woman and the infant until they disappeared from my view. Such is life in India!

A short time later, I entered the airport. Here I was greeted by a long queue at the immigration gates. Security had tightened following the 2008 Mumbai terror attacks. My camera case was thoroughly examined, and then its contents were emptied onto a tray as a security officer interrogated me. After a few tense moments, I was allowed to enter the boarding area, and off I went to Kerala!

As the plane took to the skies, puffs of white cotton balls blanketed the atmosphere and the sun's golden rays pierced through the window. I closed my eyes, soaking up the warmth and breathing deeply. Within forty-five minutes, I was awakened by a turbulence as the plane speared through the clouds and then dropped suddenly. The seatbelt signs came on and the flight attendant announced that the descent had begun. As I looked down, a dense layer of smog blanketed the landscape, hiding Kerala's lush green mountains. I wondered how many elephants were wandering below in that very moment. But as the plane got closer to the runway, I realized that they were palm and rubber plantations and vast paddy fields.

Fifty-five per cent of India's population is comprised of farmers who depend on agriculture to survive. They live on the forest peripheries for easy access to firewood and cattle grazing. But this convenient arrangement for humans is bad news for wild animals. Elephants in India are under grave threat due to habitat degradation and fragmentation, fuelled by the exponential growth of the human population. Ongoing human encroachment into wildlife habitat is intensifying human-animal conflict. The farmers and villagers living near the forest fringes install high-voltage electrical fences illegally, which kills scores of elephants every year. Trains on railway lines that cut through key elephant habitats and migratory routes also kill numerous elephants annually.

On top of all this, a disturbing trend is emerging in India with villagers taking matters in their hands. They are planting illegal dynamites and firecrackers in fruits like coconut and pineapple to lure wild boars and other animals in order to protect their plantations or farms. And elephants also become victims of such heartless actions.

As I was ruminating on the perilous situation of these ecologically significant animals, the plane hit the tarmac, bringing me back to the present moment. I gathered my baggage and exited the airport, where Raj was awaiting my arrival. We hopped into a rental car and off we drove, along the bumpy and crowded streets. The hot, humid air was palpable. It was 6:30 p.m., and about 28 degrees Celsius (82 degrees Fahrenheit). It was hard to imagine what it would be like during midday, when the temperature surged up to 45 degrees.

There was so much to catch up on since I had seen Raj earlier that year when we had witnessed the elephant's extraction from the ditch in Wayanad. We reminisced about driving down the winding hairpin bends of Ooty through the pouring rain, the lush Wayanad forests in Kerala, the fallen giant, and the precarious rescue operation to save him.

Raj knew how sensitive I was, and he warned me not to become emotional and not to judge or misinterpret what I was going to witness in the days ahead. In this, he seemed a bit concerned about my reaction to the temple elephants we would be visiting. We checked into a hotel that Raj had reserved. As I retreated, I contemplated on my discussion with Varma and Dr Sukumar. It felt like we were sort of compatriots with our common goal—saving India's elephants. With all of this, I sensed deep down that something significant was unfolding in my life. And, although I was not used to flying blind, it felt as though a higher power was directing me forward onto the next step of the invisible footpath.

Chapter 5

The Unfolding Continues

My first day back in Kerala was all about familiarizing myself with the task at hand. Raj had arranged for a local co-ordinator from his wildlife conservation organization to join us for breakfast. Next, we were scheduled to meet a Kerala veterinarian. So, after breakfast, we promptly left for the Kerala Veterinary and Animal Sciences University, where we met Dr T. S. Rajeev briefly. He escorted us to a lecture hall where he was to give a short talk about elephants, which we had made plans beforehand to attend. As an assistant professor at the university and a practising veterinarian, Dr Rajeev had treated several elephants over the years.

Listening to his speech, I was reminded of my biology courses in college, which were packed with dense scientific theories and names, that were required to be memorized. As Dr Rajeev concluded his presentation, the next lecturer, Dr Jacob Cheeran, a world-renowned elephant veterinarian from Kerala, took to the podium. Dr Cheeran had been one of the first vets trained in tranquillizing elephants. He was the keynote speaker at that presentation. Soon after his lecture, we were introduced, and I shared my interest in elephants. Little did I know then that he would play a significant role in my future mission!

Dr Rajeev then offered to take us to a nearby temple where an elephant was being paraded. As we drove through the hot and humid weather and the chaotic traffic, I could hear the boisterous sounds of drums, horns, and pipes in the distance. A few meters away, I saw scores of people lined up on a narrow street to observe the spectacle that we would momentarily encounter. In the next instant, I was startled by a beam of light reflected from a golden caparison adorned on an enormous dark grey figure. As we approached, I realized that it was a shackled bull elephant decorated with ornaments—his magnificent, large tusks clearly on display.

Childhood memories flashed through my mind and I felt a sense of déjà vu. I recalled the happy experiences mentioned earlier in this book—those of my early interactions with the majestic bull elephant in the temple that my grandparents had brought me to in the little village of my birth, Alathur, in Palakkad district. At three years old, I would happily play with this wonderful giant, unmonitored.

When I became a teenager, my grandmother told me that, as a child, I'd asked her why the elephant at our temple was in shackles, but I was not. The next day, she bought me anklets, and as she wrapped them around my ankles, she chuckled, 'Now you also have anklets like the elephant.' After saying this, she had suddenly choked back tears and continued, 'But you weren't satisfied with the anklets, Sangita. At four years old, you wanted to know why the elephant's front legs were shackled together, but your legs were not. I became speechless. Even as a child, you asked deep questions.'

Four decades later, I was still probing the same question as to why these elephants were chained. Reflecting on this childhood incident, I couldn't help but feel that some early seed about elephant abuse had been planted in my psyche way back then. Perhaps it would come to full flowering through the work I was meant to do. I pondered this as Raj and I travelled to our destination. I also wondered if this was another step on the invisible footpath becoming increasingly visible before me.

My thoughts were interrupted as we had arrived at our destination. As soon as we parked the car, I mounted my camera on its tripod, ignoring the oncoming traffic. Brushing past the crowd, I began to film this majestic elephant from the opposite side of the street. The first thing that alarmed me as I looked through my camera lens was the sight of a man holding an oil lamp so close to the pachyderm that his trunk could get burned with the slightest movement. I looked askance at Raj, and he told me that captive elephants are accustomed to these types of lamps. I tilted my camera down, zooming in on the elephant's front two feet that were shackled like those of a handcuffed prisoner. Indeed, they were similar to the locked shackles that had been etched in my mind since childhood. I then panned my camera on a long and heavy chain that was tossed over his body, linked to another set of shackles on his hind feet.

I crossed the street to take a closer look at his rear left ankle. It was utterly devastating! Several heavy chains were wrapped around to weigh his foot down and limit his movements. His rear right ankle revealed rings of depigmented skin and a number of scars on the foot, no doubt, from previous wounds. The handlers had tried to cleverly camouflage these scars with concealers and colourful anklets. The poor animal continued to shift his body from side to side, with tears flowing down his face. Despite this, a large plaque of a deity was mounted on his back and three men scrambled atop the elephant to sit on his protruding spine. Two handlers stood next to his feet, poking the animal with sharp weapons. They forced him to stand steadily and controlled his already restricted movements, making it impossible for him to evenly shift his weight.

The selfish nature of the two handlers was evident. They sought shelter from the scorching sun beneath the elephant and sipped water as the defenceless animal struggled to cope with the merciless heat. No water was offered to the elephant to relieve his thirst, nor was he given proper food, although they did proffer a few dry branches of Caryota palm.

As I was gathering camera shots from various angles, I heard Dr Rajeev's phone ring. He then signalled that we had to leave, but I did not budge. Raj walked over to where I stood and insisted that we should go. He explained that Dr Rajeev had to tend to a young elephant in his annual musth cycle and then conduct another routine check on a senior elephant. Raj and I were allowed to accompany him so that we could witness him in action as a veterinarian, while observing the elephants in their captive environment.

The contrast was all too stark in the way they treated the wild elephant that had been rescued from the trench on our earlier visit to Wayanad and this captive, enslaved elephant. As we drove past the crowd, I kept turning back to look at the defenceless animal until he faded away, as did the haunting background music. I vividly remember thinking, *what on earth could they have done to this gigantic animal to suppress his power and prevent him from breaking free of those chains of pain?*

It was just past midday when we arrived at the site where the elephant in musth was tethered. The scorching sun had intensified. Temperatures hovered around forty degrees Celsius, which, by Kerala's standards, was quite mild. I kept sipping water to avoid dehydration and fatigue. I assembled my camera beneath a large mango tree, which drew a few men and young boys who had gathered around to watch. They then curiously followed me as I walked behind Raj and Dr Rajeev.

Just yards away, a magnificent bull elephant, shackled to a cement pole beneath a canopy of trees, was swaying restlessly. They called him Kalidasan, which means 'servant of the Hindu Goddess Kali'—who is often associated with destruction, but is also considered a mother figure, symbolizing maternal love. As we approached the large animal, I noticed something oozing from the side of his forehead. It was the musth fluid, which is secreted by temporal glands. Musth is an annual mating period in male elephants when their testosterone and energy levels can surge up to one hundred per cent higher than normal. They are overwhelmed by the urge to mate. This makes them dominant and aggressive.

It was hard to ignore how aroused the poor elephant was! Some local activists had informed me that the handlers routinely tossed rocks at the genitals of elephants, causing serious injuries that often resulted in death. I wondered if this bull elephant's handlers also tossed rocks at his genitals.

In the wild, elephants find release for their surging energy and testosterone levels by trekking hundreds of miles, fighting with other bulls, and mating. But alas, in captivity, they are tethered twenty-four hours a day, seven days a week for at least three to four months—nearly one-third of their entire life. They are deprived of their primal instincts to roam freely, mate, and interact with other bulls.

Kalidasan, the magnificent bull elephant who was before us now, was just another sad tale. His right ankle was shackled severely in cruelly short chains, layered in the same fashion as the first elephant we had seen that day. He dribbled urine uncontrollably and balls of dry dung on his rear legs were infested with bugs and worms. His feet were soaked in puddles of his own making.

Elephants are clean animals. They bathe in flowing waters and indulge in mud baths to cool off. When the slush dries off, it crumbles and falls, peeling away germs from their skin. They also love rubbing against trees to scratch themselves.

But Kalidasan had no chance to experience any of these natural behaviours. I stood there, helplessly watching the poor elephant as he kept bobbing his head up and down, shifting his body from side to side, and extending his trunk as though eliciting some measure of relief from me.

Suddenly, four men arrived—two handlers and two temple authorities who 'owned' the elephant. They brought a couple of ripe, green coconuts and tossed them toward the bull from a distance, which he neatly intercepted with his trunk. He cleverly used his front foot to quash the coconut, rip it open, peel off the soft white pulp with the tip of his trunk, and then toss it into his mouth. He used the palm branches lying next to his feet to scratch his back. He then placed them on his forehead to cool off from the merciless sun. These brilliant animals are resourceful and exceedingly clever at creating tools with branches to protect and defend themselves.

I filmed Kalidasan from all possible angles. In the end, when I positioned my camera on the ground to gain some perspective on his massive size, he extended his trunk and touched it. He seemed desperate for a companion and some compassion. All of his natural instincts were cruelly denied by puny humans who could be crushed by him in a matter of seconds if he only reclaimed his strength. But unfortunately, he had been terrorized using barbaric physical and psychological torture, thereby conditioned to 'obey' human commands.

It was now time for Dr Rajeev's next visit to the world-renowned Paramekkavu Temple. Sadly, I had to leave Kalidasan behind. During the fifteen-minute drive to the temple, I kept obsessively ruminating on the poor elephant, the first one I had ever witnessed in musth. Reaching the temple and disembarking, we approached an elephant who was surrounded by approximately ten men, including four veterinarians. This elephant looked to be in bad physical shape, with protruding bones and sagging skin. Dr Rajeev explained to us that this senior elephant had severe arthritis, which he periodically examines. But he could only do this with the elephant sitting or lying on the ground. And it was the handler's job to see that the elephant did so.

As we approached the elephant, the temple authorities and bystanders gave this stranger an unwelcoming look. But with a smile, I politely introduced myself in the native dialect of Malayalam. It thawed the tension, and they smiled back, a bit surprised that I spoke their language. I then positioned my camera, keenly observing three vets who approached the exhausted and emaciated animal. Without a moment's hesitation, I pressed the record button on my camera and began to film.

A dark, bare-chested man with a saffron sarong and a long pole then entered the scene. Threatening the animal with the pointed metal end of the pole, the handler began yelling at the bull. The elephant instantly began urinating out of fear, which seemed to agitate the handler even more. His vocal commands to the elephant became louder and more aggressive. However, for at least twenty minutes, the elephant refused to respond to them. By now, the handler was becoming visibly angry and began to poke and prod the ankles of the elephant with the pointed metal tip of the pole. But the animal still refused to obey his commands.

As all this was happening, I continued to reposition my camera so I could more directly capture what was transpiring. I noticed through my camera lens that the handler was keeping a close watch on my movements. As though it had suddenly dawned on him that his cruelty could be exposed, he signalled Dr Rajeev and whispered something in his ear. The panicked vet came running toward me and blocked my camera. The nervousness on his face was unambiguous.

In order to avoid any conflict, I turned in the opposite direction, panning my camera away from the senior elephant. I spotted a massive tank filled with water, perhaps a bath tank for the elephants. As I continued to pan my camera, a pile of rubble, cement, and a forklift came into focus. The place looked more like a construction site, certainly not a spot where one would expect to see elephants! Within moments, another elephant, carrying Caryota palm branches, walked toward the temple for an evening ritual.

Then a third elephant, much younger than the previous two, was cruelly tethered beneath a barren tree that barely sheltered him from the sun. His entire body was caked with dirt; apparently, he had not been bathed for weeks, maybe, he was being punished for breaking a handler's arm. The expression in the eyes of this stunning elephant revealed harrowing tales of sorrow and misery as tears flowed down his face.

After almost fifteen minutes of gathering footage of these two other elephants, I panned my camera back to the senior elephant. He was still standing, and the handler's voice was becoming more frustrated. The assembled veterinarians shook their heads in disbelief, occasionally glancing at me. Surprisingly, they didn't even pause to think that the arthritic pain could have been preventing the bull from obeying their commands.

After forty-five minutes of this ordeal, they decided to call it a day. Too many unpredictable factors were at play. A foreign journalist was filming an elephant who was not responding to his handler's commands, and a group of veterinarians was trying to coerce the recalcitrant elephant. Completely exasperated, Dr Rajeev walked away from the elephant as the other vets followed him. He was disappointed at being unable to administer drugs to the ailing animal.

As we were getting ready to leave the compound, I remembered that we had brought some pineapples and bananas for this elephant. I signalled our driver to fetch the fruits. At the same time, the handler was approaching with a massive aluminium container full of cooked rice mixed with turmeric and lentils. It was suppertime for the senior elephant.

Just moments ago, the same handler had been irate, but now he was making balls of cooked food and feeding them to the elephant. What a contrast! After the elephant had finished his dinner, I asked if I could feed him some fruit. The handler took the pineapple and sliced it into pieces to make it easier for the elephant to eat. He then allowed me to feed him. As the elephant stretched his trunk and took the fruit, I felt the coarse texture of his thick skin rubbing against the back of my palm. It had been decades since I had felt this sensation. Again, my mind was instantly flooded with childhood memories of my early interactions with this marvellous species.

I spent the next day with a captive bull named Jairam, who was in his sexual peak. The musth cycle was pure hell for Jairam every year. This time around, he was tethered beneath a tree with few branches, barely enough to shade him from the scorching sun. His tethering site was filthy. Like some of the other elephants I had witnessed, he was standing on his dribbling urine and excrement. Perhaps days, or even weeks, had gone by since they had cleaned up the place. His handler hosed Jairam from a safe distance, giving him water to drink. The owner and handler then gave him balls of rice and lentils in a basket attached to a rope, which they positioned barely close enough for him to reach. He delicately scooped up food from the basket to ensure that not an ounce of his precious meal would spill.

Jairam would remain shackled in this contaminated place for the next three or four months. He was a prime candidate for foot rot and deadly diseases like tuberculosis. Most captive elephants in Kerala suffer from foot rot that leads to abscesses and thinning of their foot pads. This may further cause severe infection. As though this was not enough of a torture, the end of Jairam's musth cycle would bring him no relief.

Every single bull elephant in Kerala is forced to undergo a very cruel tradition, one that defies all the holy books. It is a secret ritual designed to break the spirit of the bull elephants and remind them that their masters are in control. This ritual, called *Katti Adikkal*, is a horrifying custom driven solely by a superstition that the bulls may have forgotten their commands during musth. It is similar to a ritual in other Asian countries called *Phajaan*, which involves the tying up and beating the animal by groups of drunken men. The elephant is viciously tortured continuously for forty-eight to seventy-two hours or until the men are convinced that the animal has been subdued. They do this in order to crush the elephant's spirit so that their commands are obeyed. After this ruthless treatment, the bulls become like zombies, resigned to their pathetic condition.

The only difference between Phajaan and Katti Adikkal is that the latter takes place every single year and mostly bull elephants undergo this ruthless ritual. But regardless of the semantics used to describe these practices, the underlying mission is to crush the spirit of the elephant and instil mortal fear in the animal's mind. Ironically, the handlers inflict such barbaric torture because, in a very primal way, they are terrified of the inestimable strength of the elephants. Pathetically, this is their way of masking their own fears and being dominant.

I visited Jairam four months later. He had emerged from his musth. His body was filled with ghastly wounds that had been inflicted on him in the most sensitive of areas: his ankles, feet, the patch of skin above the tusks, and trunk. There was no doubt in my mind that he had been subjected to Katti Adikkal, and I wanted to record his condition.

I quietly turned on my camera, which did not go unnoticed by his handlers. They quickly began to apply a black paste to conceal the elephant's wounds, explaining that it was an Ayurvedic cream, even as I continued to film. At the same time, I noted that Jairam's shackles were digging into his flesh. However, the handlers refused to loosen them. They then offered to bathe Jairam in a nearby lake for the benefit of my camera. When we reached the site, however, we found that the lake had been transformed into a paddy field. This made it abundantly clear to me that the handlers had not even visited the 'lake' in a long time. This also telegraphed that Jairam had been deprived of bathing in free-flowing water on a regular basis.

But the plight of the handlers is equally pathetic as well. Four to five of them share a small room inside the elephant owner's compound. They are allowed to visit their families only once a month. Many of these men became handlers because their forefathers were also elephant caretakers, thus, inheriting the family tradition. Others are young men who appear to want to show off their masculinity by controlling and torturing a massive animal. Most of the handlers live below the poverty line and try to conceal their suffering through substance abuse, which is generally country liquor.

My encounters with these four elephants revealed the truth behind all the glitz and glamour of Kerala's festivals and in so doing, split open my heart. Seeing these majestic animals enslaved and shackled like prisoners was horrifying. Ignorance and denial perpetrated the suffering of elephants. How could they not realize that these animals are incredibly sensitive? The most ruthless kind of cruelty is to abuse a defenceless animal.

As we drove back to the hotel, I knew that it was time to take a step back to see through the lens of objectivity to understand the root cause of what seemed to be an identity crisis among the youth and men who inflicted atrocities on such noble animals.

The emotions that were coursing through me had to go somewhere and a nasty argument with Raj ensued. I began swearing at him for not doing enough to end the elephants' suffering. The tension in the car was palpable but Raj didn't engage with me too much verbally in our driver's presence. However, later that evening, he threatened to leave. But I couldn't care less. Haunted by everything I'd seen that day, I was unable to sleep at all that night.

On some deeper level, of course, I was able to relate to what was going on, given my own shackled and restricted upbringing in the macho, patriarchal culture of India. I had a sense that these parallels between me and the elephants existed. And yet, it would not be until I was at the apex of my elephant conservation work that I would come to comprehend the true significance of my motivation for helping these powerful but enslaved animals.

After returning to Toronto from Kerala, I tried for weeks to release my anger and resentment towards people who only seemed to care about instant gratification. I had to come to grips with the harsh reality that a culture, which outwardly condemns cruelty, was actually exacerbating the elephants' agony.

The central and state government authorities in India deemed it too risky to enter into any discourse about the religious and cultural implications involving elephants, or any animal for that matter. They would dare not ruffle feathers, lest they become unpopular and lose political capital. Even speaking out against the cruelty inflicted upon elephants would be a political suicide, to say nothing of the consequences of actually banning them from temples and festivals. Although a few animal rights activists have been constantly sounding the alarm bells, the vast majority of people in Kerala are simply unable to comprehend the idea of cultural festivals without elephants.

Despite knowing the truth, these officials—who could wield enormous clout if they so choose—instead remain silent because they covet highly desired seats on the executive boards of the temple committees and are more concerned about maintaining the status quo. In fact, I have met highly educated professors and intellectuals who are really good at analyzing and justifying this situation. Ironically, despite achieving high academic accolades and religious beliefs, they seem to have lost their conscience. Meanwhile, temple priests have also surrendered to the commercial exploitation of elephants in order to claim their piece of the economic pie.

In so doing, each and every major religion is ignoring the basic tenets that preach love and compassion for all sentient beings. Knowing deep down that the core principles of religion are being misinterpreted, they nevertheless continue to cling blindly to cultural and social myths. Of the few people who have dared to speak out, their voices have been quickly rebuffed. Others refuse to acknowledge the grim reality of the situation because, if they did, the temple devotees would stop sponsoring elephants, and the temples would lose their operating revenue. The nexus of commerce and religion is so deeply entangled that an outsider would have a difficult time understanding the actual mechanics of this interaction between the elephants and the temple owners. Well, here's how it works.

Basically, in the same way that cars are leased out, elephants are also rented out by those who 'own' them. This is done through a middleman who usually negotiates deals between the various temple authorities and the elephant owners. He will quote the highest possible price, depending on the popularity of the elephant. In doing so, his own commission will be that much greater. The more money he can secure for 'leasing' the elephants, the more money everyone will make, including the temples.

Once the temple knows the price of the elephant, they relay this to the temple devotees. A pool of these devotees will then see to it that they collectively contribute enough money to 'rent' the elephant for the temple ceremony. In exchange, the devotees have their names published in all of the temple brochures and advertisements, giving them a sense of exaggerated status. They don't even stop to think of the agony that elephants endure in order to be displayed in festivals throughout the day beneath the scorching sun, deprived of the most basic necessities of food, water, and rest.

Here is the bottom line—temples, churches, and mosques have all become business enterprises rather than a safe haven for people to worship and find true peace of mind. Politics is rife with narcissists who apparently only care about their respective personal agendas and would go to any length to silence the critics. Power and money have corrupted even the most well-intentioned people. In today's world, everything is for sale, including elephants.

But now I knew more than ever that I had to shed light on these issues, using my media skills and background in biology. Spirit, the universe, or God—whatever name you want to call the power which is greater than ourselves—was making me see that I had been placed in a certain pivotal position in order to try to bring justice to the situation. And it was becoming clearer to me that my fate was entangled with the fate of the captive elephants of India and that there was a healing road that we were meant to travel together.

Chapter 6

MEETING MY SOULMATE

During my December 2013 visit to Kerala, I'd had an interesting encounter with a festival organizer named C. A. Menon. As an insurance broker, he had made significant connections in the community before transitioning into the cultural festival industry. He also had owned different elephants in his earlier days. However, by 2005, after all his elephants had died, he seemingly realized the foolishness of such cultural practices and became an advocate of reform instead.

He had invited me and Raj to his place in the suburbs of Thrissur district to discuss how he might provide us with access to some of the captive elephants of Kerala. As we passed through the black gates of C. A. Menon's bungalow, two elephants made out of black granite stone greeted us. On the verandah, we found a bare-chested man in a white sarong, resting in an easy chair, his legs stretched on a stool. Apparently, Menon was catching a noon siesta after lunch.

His wife came to the verandah to greet us and then ushered us inside. The first thing that caught my attention in the living room was a bull hook hanging on the wall. As I was examining the nasty iron hook, Mr Menon walked in. He explained that this weapon of torture was used to discipline the elephants he had owned a long time ago. I curiously tapped the tip of the bull hook on my palm the way they hit the elephants. And let me tell you, the shooting pain was excruciating. One can hardly imagine the agony that the poor elephants must have endured. For the record, the elephants would often receive wicked whacks on the head with the pointed iron hook. The vicious weapon is also dug into their sensitive ears and pulled, often ripping the ears' tender skin.

Menon bragged about how much he had loved his elephants. And surprisingly, he also openly expressed that he was torn between parading elephants in cultural festivals and allowing them to roam shackle-free because of the unnecessary suffering they have to endure. The problem was, he seemed to realize that it was wrong to exploit them but continued to do it anyway as it was a cultural norm. If he spoke out against the social traditions, he feared being alienated by his colleagues who were aligned with the status quo.

Menon was the first person I met in the 'elephant industry'. And with us, he generously shared the names and contact information of some of the other influential individuals. He then offered to take us to a temple where fifteen elephants were to be paraded the next day. The festival season in Kerala had just begun; it usually runs between December and May. During this period, there are at least three thousand festivals, most of them displaying elephants.

By the time we arrived at a bungalow near the foot of the temple, it was midday. Menon entered the house authoritatively and invited us along as the family inside greeted us warmly. Realizing that we hadn't had lunch, the women threw themselves into a frenzy and literally cooked up a storm, serving us a delicious meal in just half an hour. Every single dish was authentic. My absolute favourite was Aviyal, a stew made of mixed vegetables, garnished with fresh ground coconut and green chillies and seasoned with curry leaves in coconut oil. The rice pudding was also quite decadent, and the food went down well. I was now eager to get to our temple destination before the place became packed so that I could secure the best vantage point for filming. I was equipped with a mini camera and many memory cards to document as many proceedings as I could.

Right across from the bungalow was a makeshift tent adorned with palm leaves. The ground had been cemented with a mixture containing cow dung. An oil lamp was lit in the centre of a mandala that had been created with powdered grains and spices. Canopies of chrysanthemum and mango leaves, considered auspicious in Hindu tradition, decorated the long and inclined road leading from the bungalow to the temple. I suddenly heard distant sounds of drums and horns reverberating and Menon explained that the procession was getting closer.

Hundreds of people had lined up on the roadside to witness the procession led by eighteen elephants. Most women were dressed in traditional clothes and jewellery, whereas I was wearing a pair of casual black pants and a short dark blue top. As I began to set up my camera, people became curious and surrounded me. Without getting distracted, I carried on with my equipment set up.

Soon enough, one by one, the elephants—walking the hot tar road—arrived at the foot of the temple. They then climbed the steep cement walkway and entered the tall gates, lining up on the right side of the temple to be decorated in heavy, golden caparisons from their forehead halfway to their trunk. Heavy brass chains attached with ten to twelve large brass bells and chrysanthemum garlands were then tossed around their necks.

As I walked across the arena, I noticed that many elephants had deep wounds on their ankles, similar to what I had seen on the first elephant of our visit to Kerala. I also spotted a blind elephant who was being forced to perform. I hurriedly shot as much footage as I could and then moved near the temple altar. Here, using the bull hook to terrify them, the ornate elephants were being forced to bow before the altar, stretching their front legs, allowing the men to mount a heavy metal plaque—illustrated with an image of the temple's deity—on their backs. These men then climbed on the back of the elephants holding the plaques and anchored the massive structure on the animal's neck. After this, the elephants were forced to circle around the temple three times, carrying the four men, the heavy plaque, and massive chains. This was visibly weighing them down.

It was quite humid, and the scorching sun shone directly upon the elephants. They kept fanning their ears, unable to bear the intense heat. It was deeply disturbing to witness many elephants with massive tumours on their hips, their hind legs shackled together, and a heavy chain tossed on their backs. The handlers on the ground constantly poked and prodded the helpless animals with vicious bull hooks, the likes of which I had seen at Menon's place.

After dusk settled in, we shifted to an elevated platform that had been constructed above the temple arena. At 8:00 p.m., obnoxious fireworks and torches were lit in close proximity to the elephants. I could see their bodies shudder in terror at all the noise. The mayhem continued for at least five hours after which the elephants were dispersed one by one. Finally, they could eat and rest after a full day of deprivation.

As we left the arena, the haunting sounds of shackles and the deafening music of the drums and horns lingered. I arose at around 4:00 a.m. and found some comfort in my daily routine of yoga and meditation. And after a quick shower, I went straight down to the restaurant for breakfast with Raj and our team.

We then drove back to Menon's place and interviewed him on his lawn about what we had witnessed the previous day. Despite being a festival organizer, he did not hesitate to give me an honest interview. He conceded that elephants were being tortured, deprived of their basic needs, and forced to do everything they hated, including the fire. He said, 'This animal doesn't like fire at all. If there is a fire anywhere, it will run away from there. You are not allowed to run as you are chained and the fireworks are taking place very close by.' At the end of a long interview, Menon asked me to contact a popular elephant activist and insisted that we visit the Guruvayur Captive Elephant Sanctuary. He said that fifty-nine elephants were being housed in the most appalling conditions there.

Menon then invited us into his home for lunch and, later, took us to see a female elephant named Lakshmi. On the way, he explained that she had been 'donated' to a local temple by a wealthy man. As we drove into the grounds of a nearby mansion—a private residence—to meet Lakshmi, our little group was greeted by two German shepherds. They snarled and barked at us aggressively from a large kennel. Ignoring them, we walked through the massive yard. As soon as the members of my small team became engaged with the elephant's owner and handlers, I quietly sneaked to the compound's rear side.

There she was—a magnificent elephant! Lakshmi first stretched out her trunk to sense my presence, her body swaying in distress. Her trainer was nowhere to be seen, so I maintained a safe distance and silently observed her. From the moment I first stepped into her sphere of captivity, it was love at first sight. Lakshmi captured my heart and soul with her gorgeous honey-brown eyes, grace, and elegance.

I was close enough to scan her body, a habit I had developed after encountering so many abused elephants with serious wounds. It was a pleasant surprise to see Lakshmi's spotless and smooth dark grey skin. Neither were her ankles discoloured nor was there any visible injuries. The shackles on her feet were not as tight as the ones fastened on the ankles of the male elephants we had seen in the festivals. Overall, she seemed well kept and in good health, although she was lonely and confined to a small space.

As I was admiring Lakshmi, her trainer, or mahout as they are called in India, walked towards the elephant. He paid me no mind, so I quietly and cautiously followed him. I then stood silently before Lakshmi to allow her to become comfortable with me. It felt entirely natural standing face to face with this majestic animal just a couple of feet away, gazing into her eyes as she looked straight into mine. We were feeling each other's presence. When the mahout realized that Lakshmi was relaxed in my presence, he stepped back, honouring our space. It was just me, Lakshmi, and pure silence.

I became bold enough to move closer to her and rub my body against hers. She responded by rubbing hers against mine. Lakshmi seemed to feel my love and began to caress my arms with her trunk. I could sense her warm breath as she sniffed my skin with her tender nostrils. I also wiped off the tears steadily streaming down her cheeks and then rubbed her chest while standing right next to her forelegs. Lakshmi swiftly brought her trunk between her legs and touched my arm, as though reciprocating my love.

The sun was intensifying by the minute, and I realized that there was no roof over her head. Instead, a transparent green cloth had been tied to four trees in as many corners to create a canopy above her and shield her head. As a biologist, I had learnt that these animals don't have sweat glands, so they cannot cool down by sweating as humans do. This is why they constantly seek shelter from the sun and immerse themselves in water or roll in mud pools.

In captivity, the only functional way to cool them off is to hose them down. Thus, I found a garden hose and began to rinse her with it. The minute the cool water descended on Lakshmi's head she released a heavy sigh of relief. I then hosed her neck and finally her body. She closed her eyes, taking in all my tender loving care. Now it was feeding time. I had brought special treats for her—pineapples, apples, watermelons, and bananas. My driver brought a sack full of fruits from our car and dropped it next to Lakshmi's shed. She sniffed out the goodies, stretching out her trunk to welcome this treat.

I grabbed a pineapple and approached her. As I did so, she gently curled up her trunk and opened her mouth, allowing me to place the fruit inside. She then touched me and comforted me, and we bonded again on a deep level. I looked into her eyes and felt the overflowing love between us. Both of us felt a sense of belonging. She trusted me in a short time. I was in seventh heaven, sharing her space, profoundly touched by her innocence and purity.

As I was savouring every single moment of this magnificent dream, the mahout returned, announcing it was time for Lakshmi to go to the temple for her daily rituals. He controlled her every step and even slapped her face for no reason. Noticing my shocked look, he justified his actions by saying, 'She needs to be constantly reminded that I'm in control, so she will behave.' It was hard to understand why anyone would hurt such a gentle soul.

Two hours had gone by so quickly and now it was time to do the hardest thing—say goodbye to my new and trusting friend, my sweet soulmate. With a heavy heart, I watched her walk away. The haunting sounds of her shackles lingered in the air even after she had disappeared around the corner.

Apart from her morning and evening rituals, Lakshmi remained tethered in the same spot for most of the day. Even though her handler claimed that she was happy, it is hard to imagine a wild animal 'happily tethered' after having been captured and torn from her family at a young age. Granted, she did not have to work as hard as the male elephants that were exploited in the festivals. But still, it was hard to comprehend a life of boredom for an intelligent animal that would otherwise be making forays in the jungle and socializing with her herd. Such is the saga of captive elephants!

That night, the sounds of Lakshmi's shackles continued to haunt me. Here again, I began to draw parallels between the life of a captive elephant and my own. The mahout said that he loved Lakshmi but had to occasionally beat her to keep her in check. My parents wanted me to be a well-behaved Indian girl, and they would do whatever it took to mould me into their version of that. It's a cultural norm in India to express love by beating and controlling, and when my father beat me, my mother justified that he did so because he loved me.

My relationship with my father had been a complicated one. He had suffered his share of miseries and cultural conditioning. Growing up in the city of Coimbatore in the southern Indian state of Tamil Nadu, he was orphaned at the tender age of twelve. Upon his father's death, the thriving family businesses—hotels, media outlets, and banks—collapsed overnight when some of the closest and most trusted members of our rich, extended family looted every last penny.

This state of bankruptcy forced my father to become the breadwinner for his younger brother, his widowed aunt, and her son, while his friends played happily and enjoyed an authentic childhood. He joined the Indian army and started working as a stock boy, handling ration supplies. After having lived a privileged life, it must have been emotionally and physically traumatic to work day and night doing physical labour.

When he was twenty-seven, he married my mother. My parents had a very unconventional wedding even by 1950s standards when arranged marriages were the norm. Apparently, my mom insisted that she would marry no one else but my father, despite objections from her parents, as they felt my mother was too educated for my father. She had completed high school whereas my father had been unable to do so due to his family circumstances.

Even after my parents got married, my father had remained in the military where he enjoyed a very satisfying career. But one day while on duty, he got into a motorcycle accident and sustained serious head injuries. Despite receiving the best medical care in the military hospital, my father's brain injuries never healed. He suffered from intense migraines frequently. But he cleverly masked his suffering by continuing to work and kept himself busy, instead of seeking further medical attention for his condition. There was no doubt in my mind that he suffered from undiagnosed Post Traumatic Stress Disorder, or PTSD, resulting from the brain injury. Captive elephants also suffer from PTSD, their wounds intensified by the psychological trauma of being torn from their families and imprisoned day and night, deprived of the basic freedom that all living beings deserve.

Although my father was deemed to be physically fit, he was relieved of the military job that he had been so proud of. He was devastated, being unable to be the man of the family and its provider. Over the years, he gradually turned into an angry and resentful man, most often his anger erupting like a ferocious volcano and spilling over me, my brother, and my mother.

Our mother suffered silently when he lashed out at her. My brother and I, on the other hand, were petrified by it. We were also terrified of getting bad marks in school or doing anything that might contradict our father's wishes. On the occasions when he beat us, our mother would defend him by saying that he had the best of intentions and only wanted the best for us. I have no idea how she rationalized his violent behaviour.

Being a sensitive young girl, I was really afraid of my father. My fear coloured and clouded my feelings for him, and my intimidated child's mind could not see more deeply into the man. It's only now, as an adult, that I have insights into the tragic forces that shaped his life. And although I don't condone my father's violence, I genuinely believe that he loved me and his family.

I still remember how he habitually brought home neatly wrapped sandwiches made of strawberry jelly—part of the breakfast served at his job—so his children wouldn't go hungry. He worked multiple shifts so he could save money for my wedding and be able to send his kids to university. This remarkable man made many sacrifices to make me the person I am today. If my father hadn't loved me, he wouldn't have sacrificed his breakfast. Nor would he have worked so hard to give me the education and material comforts that I was able to enjoy.

He couldn't tolerate anyone saying anything bad about his family. Having worked in the military, he was a strict disciplinarian, and corporal punishment was his way of expressing his love for us. My uncle always told me that prior to the motorcycle accident, my father used to be very peaceful and tolerant. In fact, through all the unexpected twists and turns during the collapse of the family dynasty, my father had remained unfazed. He was determined to do everything possible to support his family. He pursued his duty—what he called dharma—to ensure his family's welfare.

As I reflected on my father and his character, I could see how much my dynamics with him was similar to that of the mahouts and their elephants. I knew that my father loved me, but he just didn't know how to express it in a loving manner. He never punished me to hurt me. He wanted me to be a perfect Indian girl, but the way he treated me was detrimental to the development of my confidence and self-esteem. I lost my sense of identity and, as a result, I had been unable to tap into my true potential for decades.

In the same way, the captive elephants are forced to surrender after their spirits are broken. Adding to it is the emotional torment following the separation from their family, which makes them give up on life and lose their sense of identity. They do whatever they are commanded to do in order to avoid pain and torture and to try and win compassion and kindness.

All of this was starkly at the forefront of my thoughts when Raj and I met with Dr Panicker. He was a vet and had created a history museum on captive elephants. It displayed the instruments of torture that had been used over the centuries to capture and tame wild elephants and to keep the unruly ones in line. Many of the weapons displayed in the museum had been banned in Kerala, yet, I had seen them at the festivals we attended. Clearly, they were still being used illegally. A chain made of a heavy iron ring with spikes pointing inward was the most vicious of them all. It is used to capture elephants who run amok and those who 'misbehave'. The spikes dig into the flesh through the elephants' sensitive skin, causing unimaginable pain and suffering. Dr Panicker also confirmed what I'd already heard before. He explained how mahouts threw rocks at the genitals when bull elephants come into their musth cycle. This inhumane and ruthless treatment has killed many elephants in Kerala.

As I began to film the weapons, Raj tried to dissuade me from doing so. He seemed more concerned that I would expose the truth, asserting that my interview questions were intentionally structured to highlight the negatives. The fact is, I was only trying to get to the root of the problem. Unable to bear his constant nagging and presence, I declared that the truth would not be silenced. Raj left the museum upset, and I realized that we were getting close to a final parting of the ways. Despite our opposite worldviews, Raj and I managed to be civil to each other in the little time that was left.

After leaving the museum, he and I, and our small team of associates deliberated what our next stop would be. You may recall that C. A. Menon had insisted that we visit the Guruvayur Captive Elephant Sanctuary, also known as Punnathur Kotta. Although this destination wasn't included in our original travel itinerary, I suggested that we go there next. But my team argued that it would be better to visit the place the following day as we would be exiting the state from that same area on our drive to Bangalore. I initially agreed with them because the logistics made sense. Yet, my nagging intuition—that we should visit the temple sooner rather than later—lingered. After breakfast, I insisted that we go to Guruvayur that afternoon, and my team reluctantly agreed.

We arrived at the Guruvayur Captive Elephant Sanctuary at around 1:00 p.m. The first thing we saw when we reached was a sign that read 'Punnathur Kotta'. *Kotta* means fort in the native tongue. Just a few feet away from the entrance, a lonely male elephant was tethered to a tree, struggling to cope with the heat of the scorching sun—a scene we had seen all too frequently before. He was also scratching his back with a palm branch held in his trunk. I left the tripod in the car to remain as inconspicuous as possible. There was no one around, but you never know who might suddenly appear on the scene. With my handheld video camera, I began to record him.

As I took a few steps toward him, Raj pulled me back to ensure that I maintained a safe distance. He pointed me to the sanctuary's entrance. Later, he revealed that he had received a couple of threatening phone calls. Apparently, some of the festival staff had been forewarned that I was documenting the plight of captive elephants. Raj was concerned that if I wandered off unmonitored, I could end up in real trouble.

We paid our admission and camera fees and walked inside the so-called 'sanctuary'. Inside, vendors were selling balloons, toys, and statues of Lord Guruvayurappan who is believed to be the reincarnation of Lord Krishna. We then entered an area where mostly bull elephants were tethered just yards away from each other. Their dark eyes were orbs of misery, oozing tears, and revealing a sad history of barbaric torture. It was evident that their innocence had been oppressed. They looked like the concentration camp victims.

Signs bearing directives like 'Elephant in Musth, Keep Distance' were plastered in front of most of these bulls who displayed signs of musth. After having witnessed the heartbreaking plight of Kalidasan and Jairam in their musth, here, I was surrounded by fifty-nine similarly tormented elephants. Most of them had musth fluid oozing out of their temples and urine constantly dripping on their own feet. These elephants were forced to stand in their excrement and urine throughout their musth cycle which lasted for three to four months a year.

The sight of one elephant was particularly disturbing. He was swaying back and forth, in a desperate weaving motion, trying to break his chains. You don't have to be a rocket scientist to figure out that he was in distress. I gathered a few shots of this elephant. As I continued to explore the area, I spotted more elephants in worse conditions. On top of this, there was no water in the vicinity to quench their thirst in the forty-degree-Celsius heat.

In the wild, elephants drink at least 150 to 200 litres of water a day. But here, they were denied this basic necessity of life. Elephants also have an incredible sense of taste and they love a wide variety of food. Indeed, in the wild, they tend to eat as many as 200 varieties of branches, roots, fruits, leaves, bark, shrubs, herbs, grass, and even soil to meet their basic nutritional needs. But in this so-called elephant sanctuary, they munched on fibrous Caryota palm branches all day long.

In 2018, thirty-four captive elephants had died in Kerala; in 2017, twenty captive elephants died before their prime; in 2016, this number was twenty-six. Eighty per cent of these eighty elephants reportedly suffered from an ailment, locally called *erandakettu*. It is an impaction of the colon and causes severe constipation, due in part by a diet consisting only of the fibrous palm leaves. But as noted above, they are meant to eat a wide variety of greens.

Worst of all was the fact that these elephants had no roof to shelter them from the scorching sun. There were a few sheds made out of tin, but they only intensified the heat. The inhumane deprivation of the elephants' basic necessities of life was all too obvious in this 'sanctuary'. One of the most pathetic sights was that of an elephant who had to stretch forward desperately for just a sip of water from a tiny cement tank on the ground. So short were his shackles. I could tell even from where I was standing that the water was filthy. I had noticed plastic debris floating in its tank. Not too far from him, another elephant was struggling to stand properly. One of his forelegs had been tethered with such cruelly short shackles that he was unable to place his foot on the ground. Deep cuts and infected ankles resulting from heavy and rusted chains, and rings of depigmented skin from chain injuries were common features in all of these elephants.

As we circled back to the entrance, I found myself in front of the same distressed elephant I had initially encountered when we'd entered the compound. I sat on the ground, just a yard away from him, to film him from unique angles. Suddenly, I was showered with litter. I looked up to realize that, out of sheer desperation, the elephant was tossing clumps of mud at me. What on earth could have triggered him to do that? If only I could have communicated that I was there to help him!

Raj, who was watching this scene unfold, warned me that we were putting ourselves in jeopardy and that we should now leave. As we made our way to our car, the sights and sounds of this nightmarish elephant factory farm kept replaying in my mind. Through the use of barbaric torture, their spirits had been crushed and subdued forever. Given that elephants never forget, they must be going crazy thinking about their past life, the peaceful wilderness where they roamed freely and lived harmoniously with other animals.

The parallels between the treatment of these elephants and the treatment of women growing up in India was also crystallizing. I was beginning to connect the dots . . . and this issue has been elaborated in one of the ensuing chapters. No wonder I had been chosen to document the captive elephants' plight.

Suddenly, my phone rang. It was C. A. Menon on the other end, checking to make sure that we had visited the Punnathoor Kotta. He informed us that our batch of visitors was the last for the year as the temple authorities had closed the sanctuary to the public. He went on to explain that the elephants were in their musth cycle, making it too dangerous for visitors. With a slight chuckle in his voice, he said that we were fortunate to have made it there just in time and applauded my decision to visit the temple that day. I breathed a huge sigh of relief and glanced at Raj with a wide grin. He smiled sheepishly back at me. He, too, realized that had we arrived the next day, as originally planned, we could not have filmed the elephants inside.

While connecting the dots, I sensed that these occurrences were not mere coincidences. Had I ignored my intuitive nudge and visited Guruvayur the next day—as our team had suggested—we would have missed the opportunity to gather key evidence of India's heritage animal being tortured and exploited for profit at a world-renowned temple. And as though it was meant to be, as soon as we had our footage, the temple officials closed off the access to the captive elephant 'sanctuary'.

In August of 2014, another synchronicity revealed itself. The Animal Welfare Board of India (AWBI) launched an investigation into the Guruvayur Temple after receiving complaints from local animal rights activists and the media. The report that AWBI subsequently put out was damaging to the temple. It precisely substantiated everything I had documented during my visit in December of the previous year. Almost every paragraph of AWBI's scathing report corroborated with the footage I had recorded.

With the AWBI's report to back up the veracity of my own work, I was able to produce a detailed segment on the Guruvayur elephants who worked tirelessly to generate hundreds of thousands of dollars for the temple. This was a temple that, despite being a UNESCO World Heritage Site and a place where all living beings were expected to be treated compassionately and humanely, had lost all its credibility.

Suddenly, the meaning latent in this series of synchronicities came into my sharp focus. By putting one foot tentatively in front of the other, I was carving out this predetermined path of my life's calling. After having watched the sufferings of Lakshmi and the other elephants, I simply could not turn my back on them. I promised these sentient beings that I would do whatever it took to expose their plight and bring an end to the atrocities that were being committed against them. By this point, I had a better idea of what I needed to do. The only question that remained was how.

Chapter 7

Learning to Let Go

I returned to Toronto with twenty-five hours of footage, not knowing what to do with it. I had neither the money nor a clear vision for my future. However, I did know one thing—I had to expose the truth about the elephant abuse I had witnessed in India. As you might imagine, this required a constant leap of faith on my part. I prayed for guidance as to how to best proceed. I had learnt the spiritual lesson that intention overrides obstacles. Thus, I prayed to Lord Ganesh, the remover of obstacles, to help me forge forward in meeting and overcoming the challenges that lay ahead.

I spoke to one of my previous editors to explore the possibility of creating short films from my footage and releasing them on YouTube. But, of course, I didn't have the money to pay him. I struggled for weeks trying to come up with possible ideas about other outlets for my footage, but for various reasons, I couldn't make this a real priority at that time.

By the time 2014 rolled on, I had dealt with prior commitments, and now, I could worry about what I might do with the footage I had shot in India. I thought about making a documentary, however, the money required for that had initially put me off. But nothing else was presenting itself. I had even come up with a tentative title for the film: *Gods in Shackles*. I kept dismissing the idea because it seemed so daunting. Not sure what to do, I continued to pray and meditate on what my next steps might be.

Heeding the advice of a wise friend, I phoned Bill Hutchison—a trusted media colleague and a popular and well-respected news anchor in Toronto. We had a friendly chat, catching up on each other's lives. At one point, I casually mentioned that I had visited India and shot more than twenty-five hours of devastating footage documenting unimaginable atrocities against elephants, and how the issue had to be exposed. I shared my idea of, maybe, trying to produce a documentary with the footage while acknowledging that I did not have the resources to do this.

Bill immediately suggested launching a crowdsourcing campaign in order to raise funding for the film. Soon after our conversation, he sent me a web link to Indiegogo, the world's first crowdsourcing enterprise. I studied Indiegogo's model thoroughly and came up with a strategy that I thought might be viable for what I was trying to accomplish. I next contacted Maneesh Malhotra, a film editor I had worked with on my Bermuda documentary series. I wanted to determine whether he would help me edit a five-minute trailer and some graphics for a nominal fee. With his help, I launched my crowdfunding campaign in early April of 2014. My goal was to raise a modest amount of forty thousand dollars. With this money, I intended to upgrade my camera equipment, hire a cameraman, and travel back to India to film the mother of all festivals, Thrissur Pooram.

There are a few grand festivals in Kerala every year for which elephants are transported in trucks from various districts across the state. But the famous Hindu festival of Thrissur Pooram is, perhaps, the grandest of the lot. It is held annually at the Vadakkunnathan Temple in Thrissur district. It commemorates the astrological event of the moon rising with the Pooram star in the month known as *Medam* in the Malayalam calendar. This usually falls towards the end of April or beginning of May. Music, fireworks, and flag-raising ceremonies are traditional features.

Sakthan Thampuran, the ruler of what used to be called

the Kingdom of Cochin (also known as Kochi), was the brainchild of Trissur Pooram. In the late 1700s, Arattupuzha Pooram used to be the largest temple festival in Kerala, wherein temples in and around the city of Thrissur participated. However, one year, they were denied entry into this festival because they arrived late. So, the participating temples lodged a complaint with Sakthan Thampuran. Thus, he created Trissur Pooram, welcoming all the temples to bring their deities and pay obeisance to Lord Shiva, the deity of the Vadakkunnathan Temple. But during those days only a few elephants were used in Trissur Pooram. In recent decades though, they have been engaging hundreds of elephants, elaborately adorned and paraded as part of the overall spectacle. In fact, Thrissur Pooram has become a major tourist attraction and is very important to the financial health of Kerala because of the annual revenue it generates.

Over the years, various animal rights activists have worked to ban the exploitation of elephants' festivals, but without much success. The documentary I was putting together about temple elephants in India would be incomplete without the inclusion of this famous temple festival and the elephants who were such an integral part of it.

The festival was still a few months away, but I was heartened by the initial response that our fundraising campaign received. Meanwhile, Hutchison and other media colleagues afforded the media coverage, which was just the icing on the cake. It boosted our campaign, and people from around the world contributed generously towards the production. My previous donors from Bermuda opened their hearts and wallets, and one man in particular, Andrew McKay, donated more than twenty thousand dollars. Several others gave more than a thousand dollars each. In the end, a total of around twenty-five thousand dollars came from Bermuda. Andrew would end up becoming our honorary executive producer, one of the modest perks bestowed upon him as a token of appreciation for his immense generosity.

Another generous soul from Hong Kong, Amanda Loke, sent me a Facebook message stating, 'I have been following your amazing work for some time now and would like to make a donation to your cause . . . Your trailer for *Gods in Shackles*! I can't get the images out of my mind . . . It makes me happy knowing I can contribute back and give them a voice. It's my way of helping even if just a little.' In honouring her wish to keep the amount of her donation confidential, I cannot share what she gave. Let's just say it was significant and she, too, received the title of honorary executive producer on the film as well.

In the meantime, I had signed up for an online social entrepreneurship course with a life coach, Ryan Eliason. I was most impressed by his love for nature and animals. Little did I know when I signed up for his programme that he would play a significant role in not only promoting my campaign to his clients but also by donating generously. Artists like Elaine Wong, Scott Charles, Crystal Wong, Ros Coleman, Ella Birt, Patricia Mastrandrea, Donna Drozda, and others contributed their artwork, jewellery, and reiki pillows for auction. Ryan also introduced me to his digital strategist who, after watching the film, decided to volunteer his time and digital skills to promote *Gods in Shackles* online.

By mid-May, we had surpassed our goal by two thousand dollars, having raised a total of just over forty-two thousand dollars! I was gobsmacked! People from all around the world joined forces to help raise the funds needed to produce *Gods in Shackles*. It suddenly became a global movement for temple elephants. This further confirmed to me that I was on the right track. If there ever was an instance of the universe supporting me and my intention to help the captive elephants of India, this was it! At this point, I felt that I could not have received any more encouraging message of validation for what I was attempting to do.

Now, it was just a matter of finding a cameraman with the right equipment for the kind of investigative footage I needed. An ideal person would be a student, eager to explore new opportunities and willing to honour cultural diversity. I was planning to advertise this position but then, all of a sudden, I was reminded of someone I had recently met—Chris Palmer, a renowned nature and wildlife film-maker as well as an author and professor at the American University (AU) in Washington State. Palmer also taught film-making and media studies at the AU, and I was sure that he would know someone suitable. He had come into my life at just the perfect time, and when I broached the subject with him, he wholeheartedly agreed to help. He went on to advertise the position at his university bulletin and, after conducting many rigorous interviews over Skype, a student by the name of Tony Azios snatched up the job.

Right from the get-go, Tony seemed organized and conveyed a positive outlook in the way he spoke about his life's mission. I explained the cultural sensitivities and controversies surrounding the project and its filming, as well as the unchartered territories that we would be venturing into. Tony took the time to conduct his own research and familiarize himself with what he was getting into. We also discussed and finalized the additional camera and lighting equipment that we needed to order. After obtaining his visa to travel to India and receiving the requisite vaccines, the two of us were all set to embark on this journey together.

I was scheduled to fly out of Toronto on 3 May 2014 to London. There, I would meet Tony for the first time. He was flying from Washington to London. And from London, he and I would take a connecting flight, together, to Mumbai. However, when I checked in at the Air Canada counter, I was told that there were no seats left. I flipped at this as I had booked my ticket two months in advance! I spoke with the Air Canada agents, explaining that I had timely work to do in India. Yet, they didn't budge. Looking into my teary eyes, one kind Air Canada agent told me that if somebody cancelled at the last minute, I would be the first in line.

Pacing back and forth, I was trying to cope with so many uncertainties thrown at me all at once. First, I was about to meet a complete stranger for the first time and establish a good working relationship with him in such a short period. Next, our appointments in India were not buttoned down. We had neither confirmation of the interviews we were planning to conduct nor the necessary permits to film the elephants being paraded in selected temples and festivals. I had no idea of what I was getting myself into in India, embarking on this very tentative journey. Everything was up in the air . . . except me!

This situation—of no seats left on the plane—was a lesson in surrender and trust presenting itself to me. Recognizing it for what it was, I told myself to just let go and let things unfold the way they were meant to. I could not force the outcome of the situation and I relinquished my ego's need to be in charge. Lo and behold, just when I had resigned to the possibility of taking another flight, my name was called out! I walked through the boarding tunnel with my head down, dragging my carry-on baggage. Even though the outcome of the situation had been favourable, my emotions were running amok. What a close call! I looked down at the T-shirt I was wearing. It featured the image of a bull elephant and I smiled as I touched his face, thanking this guardian angel of my trip for seeing that I got on the flight.

My flight arrived at Heathrow Airport in London just in time for me to get to another departure gate where I would meet Tony and from where we would depart for India. I dashed to the gate, and there he was, a handsome young man waiting to receive me with open arms, literally. Tony and I hugged as though we had known each other for a long time. Indeed, we had talked quite a bit on Skype and were fairly familiar with each other. Tony was in his mid-thirties, of average height, and with a crown of dark curly hair, dark-brown eyes, and light skin. Projecting a calm demeanour, he gently handed me a gift that was delicately wrapped. I impatiently ripped it open. It was a pair of yellow socks adorned with a grey elephant print which Tony had very thoughtfully bought for me.

Earlier, Tony had already set up his camera in the passenger lounge in his efforts to weave together a narrative of our journey. Now he began capturing relevant shots. A short time later, we were promptly ushered into our connecting flight. During our in-flight dinner, Tony explained that he was equipped with the latest technologies. Indeed, I was excited to see his camera and all of the lighting equipment that he had purchased for filming as per our discussions. He also had a camera for time-lapse videos in addition to a cool GoPro camera that could be mounted on a car to obtain unique aerial shots.

After nineteen gruelling hours, we finally arrived at the Chhatrapati Shivaji International Airport in Mumbai. It was midnight and, needless to say, we were exhausted. After grabbing a bite to eat and a much-needed jolt of java, we relaxed at the airport until about 7:00 a.m. before boarding our next flight to Kochi, Kerala. From there, we would make our way to the city of Thrissur, the cultural capital of Kerala. Here, the festival of Thrissur Pooram is held every year in which almost a hundred elephants are paraded over the course of its thirty-six-hour duration. Despite having just flown into India, we had several appointments scheduled for the day and we knew we had no time to waste.

In less than two hours, our plane was approaching a lush green Kerala. I had learnt from my previous trip in December 2013 that much of the natural habitat here had been replaced with palm, tea, and rubber plantations. I also got a bird's eye view of Kerala's well-known canals. As Tony and I exited the Kochi airport, a tall, skinny, and dark man with sharp distinct features wearing a white shirt and sarong waved at us. It was Salu, the same driver who had driven my team around in

December of 2013. He greeted me with a flashing white smile and loaded our bags in the car. He was reliable and I knew I could trust him to remain discrete about our unfolding schedule and the secret locations we planned to visit.

During the two-hour drive from Kochi to Thrissur, Tony was able to film some unique sights and sounds including trucks with paintings of Lord Ganesh on them and white-coloured churches against a stunning backdrop of dark clouds. He also filmed billboards advertising jewellery, many of which portrayed well-known actors standing beside caparisoned elephants. As we entered the city limits, the congested traffic and pollution became too much to bear. It was nearing midday. The sun's heat intensified by the minute and the hot air and soot made it difficult to breathe. The noise pollution also became unbearable, with trucks and cars honking obnoxiously. We rolled up the windows for some peace and air conditioning.

It's important to note that after my first trip to Kerala, I stayed connected with everyone I had met, including the festival organizer C. A. Menon. I had told Menon that I wanted to, at some point, visit and film the big festival, if I could raise sufficient funds to hire a cameraman. On our way to the hotel, I phoned Menon, who offered to organize our itinerary and provide us with media passes for the festival. So, we stopped at his place and spent a couple of hours scheduling interviews and discussing the best way to film the grand spectacle.

Menon also reminded me to connect with a local activist, Venkitachalam. He was a lone voice raging against the use of elephants in Kerala's festivals. I had read many media reports about this man. In fact, he'd been featured in the *New York Times*. In my research, I found out that Venkitachalam is one of a kind—a devout Brahmin who follows the core Hindu tenets of non-violence, compassion, and truth religiously. He has two purposes in life. One of them is to care for his 90-year-old mother, and he chose to remain a bachelor to be able to do so. The second is to be a voice for the captive elephants of Kerala. He has fought several court battles on their behalf, pushing the Kerala government to create new laws in the process. And despite several death threats from various elephant owners, he has never given up on the defenceless elephants.

Venkitachalam lives in the heart of the cultural hub near the festival grounds and he was our next stop. Tony and I drove along a narrow, congested lane in the city of Thrissur. I was reading our directions aloud and knew we were getting close to Venkitachalam's house. As we drove along, all of a sudden, I noticed a tall, lean man with grey hair and a beard standing on the doorstep of a blue-coloured house. Venkitachalam waved at us as we pulled over. Carrying our camera gear, we made our way up the steps to greet this kindred spirit.

Venkitachalam was in his late forties. During the day, he taught accounting to senior students, and in the evenings, he connected with his cohorts to gather information covertly on the plight of the elephants. Behind him, I noticed an elderly woman with a crown of silver-grey hair that was tied into a small bun. She wore a traditional red sari embroidered in golden thread and a warm smile that emanated through her eyes. Slightly over four feet tall with a hunched back, she had to strain her eyes to look at me. She resembled my beloved grandmother who had raised me in Kerala, and I was instantly drawn to her.

As my parents had taught me, I bowed down and touched her feet in a display of respect to elders. She quickly grabbed my shoulders and hugged me, then, in a gentle voice, asked me my name. Catching her breath, she held my arm for support. I wrapped my other arm around her shoulders, leading her to a wooden bench where the two of us sat down together. We looked at each other silently. She cupped my face with her two palms and gazed into my eyes affectionately. I could feel pure love springing from her heart.

As I was chatting with this wise and kind soul, I realized that Tony had disappeared. So, I stepped into the backyard of the house, looking for him. Native trees and plants thrived, as did a small vegetable garden where tomatoes, pumpkins, and beans hung from the creepers. Coriander and curry plants flourished in the rich, red earth. There was also a deep well that provided water for the daily needs of Venkitachalam and his mother.

I spotted Tony surveying the backyard where he had set up the camera. After clipping a wireless microphone on Venkitachalam and testing the sound levels, we began to interview him. He did not mince words, nor did he hesitate to critique the religious establishments. 'Ten festival elephants had collapsed and died since January 2014, one dying of heartbreak, terrified by a thunderbolt,' he divulged.

Venkitachalam spoke to the camera as a living encyclopaedia. He knew the exact age of various elephants, the towns they came from, the states they were illegally bought from, the names of their owners, the number of people they had killed, and so much more. He also said that heavy downpours were expected for this Thrissur Pooram and warned that many elephants may end up running amok as a result.

The interview lasted for about two hours, given that there were many interruptions caused by the detestable traffic and its noisy horns. Following it, Venkitachalam gave us a tour of his modest home. It was furnished with a couple of wooden chairs, benches that sat on the cement floor, a tiny kitchen with an ancient stove, and a couple of open shelves on the wall where traditional utensils were arranged meticulously. Next to the kitchen, a cot was draped in blankets covered with a cotton bed sheet. There was also another little room where Venkitachalam taught accounting.

There were no sofas or elaborate dining or coffee tables or even mattresses for that matter. A life of simplicity and contentment appeared to be Venkitachalam's chosen way of being in the world. Despite having so little, he and his mother offered us whatever they had with pure and generous hearts. I particularly loved the irresistible *ladoo* they proffered. It's one of my favourite Indian sweets and their love made it extra delicious!

With the day slipping away, I sat in Venkitachalam's living room and began organizing interviews with key stakeholders for the next few days. The time flew by and soon we had to go. As we were leaving, Venkitachalam's mother gifted me a brass oil lamp that I light in my shrine daily, and something I will cherish until the day I die. I hugged her and held her close to my heart for almost a minute . . . so special was our bond.

Meanwhile, thunder rolled over our heads and lightning split open the skies, making way for a heavy downpour. News about the torrential rains flooded the networks, and the weather forecast for the next three days looked gloomy. In fact, the newscasters were even predicting that the unseasonal tropical rain could lead to the cancellation of Thrissur Pooram. The uncertainties surrounding this proved to be just as stressful as boarding my flight from Toronto. After crossing the ocean, it appeared that the main event that we had travelled so far to film, may not even take place. My resilience and trust were certainly being tested again. I had to constantly remind myself to surrender to what the universe had in mind.

On the one hand, I felt unsettled by the unknown, but on the other, my heart was cheering for the elephants, for they love the rain. We drove around the city, gathering footage of the puddles, as the flooding water gushed through broken drains. We even took our chances travelling to the ocean where palm trees were being blown away by gusty winds that uprooted some trees. To witness the ferocious sea and the roaring waves lashing on the shores was a true reminder of Mother Nature's sovereignty and power.

Later that night, we checked into our hotel, had a light dinner, and retired early to bed. I was so jet-lagged that it should have been easy for me to sleep but I was struggling to deal with so many uncertainties that were constantly being thrown at us. What if the rains were so torrential that the festival got cancelled? What if the camera malfunctioned? What if the equipment got damaged by the moisture from the rain? What if this . . . what if that . . . ?

Just as panic was beginning to set in, I remembered that I would be meeting my beloved elephant, Lakshmi, the next day. All my tensions melted away immediately as precious memories flashed through my mind and put me to sleep. The next morning, I was up by 4:00 a.m. and performing my morning routine. I was ready to leave bright and early as Lakshmi usually arrived before 6:00 a.m. for her daily rituals at the Thiruvambadi Temple.

Rain or shine, rituals are conducted every day in all Kerala temples, where they mostly use bull elephants. In fact, not only does this state house the second largest number of captive elephants in India, but the largest number of captive bull elephants as well. Female elephants are a rare sight in Kerala's temples. And although Lakshmi performs the daily temple rituals, she wouldn't be allowed to participate in the cultural festivals, certainly not in the Thrissur Pooram festival. The misogynistic cultural attitudes were all too obvious even in the way they exploited elephants.

On this particular morning, I was clad in a black-and-white chequered sari with elephants imprinted on a red border, ready for my date with Lakshmi. Tony seemed a bit surprised as he had not seen me in a sari, but still, he forced a smile. We briskly made our way to the hotel's restaurant, and after our breakfast, loaded the camera gear into our car. Salu was punctual as usual and he drove us to the Thiruvambadi temple, which was about fifteen minutes from the hotel.

It turned out to be an auspicious occasion, and with the Thrissur Pooram festival just days away, the temple was jam-packed. But Lakshmi had not yet arrived. We noticed a young bull elephant inside the temple. As we approached him, he stretched out his long, elastic trunk. For the first few minutes, I observed him silently from a distance, trying to gauge his emotions. As Tony and I began to set up our cameras, all eyes were on us. Some people looked irritated while others were mesmerized by Tony's white skin.

As soon as we positioned ourselves in different corners of the temple, I heard the haunting sounds of shackles. And there she was! My beloved Lakshmi had entered the temple from the back entrance. But before taking her closer to the main deity, she had to be hosed down thoroughly. Next, her mahout gave her some water to drink and then took her to the usual tethering spot and shackled her front two feet. She stood there swaying from side to side, no doubt, going out of her mind.

When we were done filming all this, we turned our attention to the same young bull elephant we had met before Lakshmi had arrived. His shackles had been severely tightened during the short while that we'd left his site. The clever bull sniffed out the bananas that I was hiding behind my back and stretched his trunk all the way around my back to grab them. He then inserted his trunk through my sari and slurped my bare belly with his long straw of a trunk. As Tony was filming, the worshippers turned into spectators and giggled, watching the elephant flirt with me. My encounter with this young, gorgeous animal will remain etched in my mind forever.

After filming the impromptu playful scene, we returned to Lakshmi. She was about to be decorated for the daily ritual procession. This involved adorning her forehead with beautifully crafted decorative bells that hung down like a necklace and a decorative headdress known as *nettipattam*, which was tied around her neck. As her handler unfolded the large glittery medallion, Lakshmi took the cue by offering up her right foreleg, which the handler used like a ladder, to climb onto her back so that he could continue to adorn her. He then rested the heavy caparison of a headdress on Lakshmi's head and allowed it to unfurl before placing it perfectly. Next, he tied off the rope that secured it around her large ears. All the while Lakshmi stood stock still, without even flapping those huge ears. After the handler had finished decorating her, Lakshmi raised her right foreleg so that he could descend down from her back.

In a few minutes, five men carrying traditional drums and horns lined up in front of the altar and began to play the instruments. They then invited the priest and signalled Lakshmi to proceed before the altar. After she took up her position, the priest emerged, carrying a statue of Lord Krishna, the temple's deity. This was Lakshmi's signal to bend down so that the priest could then climb onto her back. Then off she went to perform her ritual duties, otherwise known as circuits, as the devotees followed her closely. After three circuits, Lakshmi returned to her spot to be handcuffed again. Here she also posed for photographs. During this time, people were allowed to feed her junk food in exchange for rewarding the handler.

At the end of a few hours of this, she would repeat the entire process, at which point it would be approximately 10:00 a.m. Lakshmi would then walk away to a more secluded area where she would trumpet three times in order to earn herself a handful of rice. I offered her bananas, and the temple devotees fed her some puffed rice and coconut, but these snacks are totally insufficient to satisfy her voracious appetite.

In the wild, elephants typically wander across vast areas for 16–18 hours a day, grazing on grass, roots, barks, leaves, berries, and even soil. But temple elephants are usually starved until after the rituals to ensure that they don't defecate or urinate during the ceremonies. In the wild, Lakshmi would be socializing with her herd. But here, she is lonely, forced to stand still for two to three hours, continually, on the hard, granite floor. Her soft cushioned feet are designed for lush jungles and marshes, not hard floors.

Again, how ironic that God's own creation, the embodiment of Lord Ganesh, was being revered and defiled at the same time! Over the years, this had become a cultural and social norm. None of the worshippers seemed to realize that these elephants had been and still were being ritually abused so they would be obedient.

The disconnect became all too conspicuous in every situation. The devotees chanted mantras of love and compassion, but they seldom practise these sacred tenets as it pertains to the treatment of temple elephants. Despite their lip service, people are unable to live by the tenets of the ancient Hindu scriptures. Such stark paradoxes lay at the heart of this dilemma.

I tend to look at the physical structure of a temple as the symbolic representation of our body. The altar is our heart and the statue of the deity inside the physical altar is the Spirit. In the Hindu tradition, we prostrate before the altar. To my mind, this epitomizes surrendering to God, which is really the Spirit that resides in our heart. The rituals and festivals are meaningless without expressing love and compassion for all living beings. In particular, the exploitation of India's heritage animal in the name of religion is the exact antithesis of everything preached in the sacred Hindu tenets.

The rest of the afternoon, Tony and I shot more footage. Given that we had started very early in the morning, at about 6:00 p.m., we decided to call it a day. After dinner, I asked Tony to bring his camera into my room so we could view the footage that had been shot thus far. I was excited to see what we had.

When we began screening, we noticed that the first shot was distorted. I realized that the camera was adjusting to the location, so I didn't make a big deal out of it. But as we kept reviewing more videos, it turned out that most of the shots were distorted and had innumerable technical glitches. The weather was too hot and humid for the camera and the images looked foggy. We could barely salvage any footage. This meant the entire day had been wasted. Naturally, I was very disappointed. But more importantly, I was worried that the temple authorities might not allow us to film again in the temple.

The mood in the room became very intense, and I asked Tony to leave. I needed some space to process the challenges that this scenario presented. I quietly meditated on the recent turn of events. I had known all along that this would be a challenging trip. And here I was, trying to deal with a big one!

Through my self-improvement books, I had learnt that matter springs from consciousness, not the other way around. And I called upon the power of intention to calm the troubled waters. Quietly to myself, I repeated my intention to expose the plight of the captive elephants of India to the world. After fifteen minutes of reciting this mantra, I was sufficiently soothed to be able to go to bed. From then on, I would include this specific intention in my meditation practice every day. As long as I held onto my intention through thick and thin, everything would be okay.

The next morning, Tony and I met for breakfast. Although I tried to suppress my irritation with him, the friction between us was palpable. I gave him the silent treatment for the first few minutes before the floodgates burst open and I shared my disappointment with him. We argued for a little while, and I thought he would walk out of the restaurant and out of my life for good. But after a few minutes of silence, and some deep breaths, cooler heads prevailed.

Having regained some objectivity about the situation, we discussed our strategy of filming the mother of all festivals the next day. Thrissur Pooram would feature almost one hundred elephants in batches of fifteen to twenty. Our plan was to capture as many of them on film as possible. What I didn't know was that the horrors that we had documented thus far with our camera would pale in comparison to what we would witness the next day.

Chapter 8

The Abomination That Is Thrissur Pooram

During the festival of Thrissur Pooram, the entire city of Thrissur comes to a standstill for almost two days. The day before the spectacle, main streets are closed for thirty-six hours as of midnight to accommodate the revellers and the elephants featured in the festival. We were advised to check into a hotel closer to the area where elephants would be paraded the next day. This was to allow us easier access to filming. Another benefit was that this hotel had a terrace from where we could gather aerial shots. Here, Tony was able to plant his slick GoPro camera. It was also an ideal place to generate time-lapse videos.

Dark, dense clouds filled the skies and it poured relentlessly. It felt as though the heavens were sounding ominous alarms to catch people's attention and stop them from exploiting the voiceless elephants. Then, at about 8:00 p.m., the rain finally subsided, allowing Tony and me to survey the grounds and plan our shots. As we stepped into the area, our feet sank into the slushy red earth, and we were not nearly as heavy as the elephants.

Not too far from us, elephants were being unloaded from trucks like commodities, taken beneath the trees, and tethered there. With no roof over their head, elephants were vulnerable to the elements. This could be disastrous, for lightning, thunder, and a heavy downpour was expected that night. As Venkitachalam had mentioned during the interview, a bull elephant had been struck and killed by a thunderbolt earlier in the year.

All too frequently, these elephants fall off from the open trucks and are irreparably injured in the process. One handler was dragging a bull elephant that looked utterly exhausted from his journey. His eyes were droopy, and his shrivelled face revealed sad tales of sleep deprivation and starvation. When I asked the handler if the elephant had been fed, he gave me a nasty look and walked away.

Ninety-five tuskers had been trucked in from far and wide to star in one of the most extravagant shows on the planet. These bull elephants had been transported from different districts across Kerala. Allegedly, this was after they had undergone a medical examination, mainly to ensure that they were not in their musth cycle. (Not being in musth was about the only criteria for festival selection.) Later, I would discover that many elephants were blind, and senior elephants, suffering from geriatric ailments, were illegally paraded.

The Thrissur Pooram focuses on two temples that are very closely situated to one another: Paramekkavu Bagavathi Temple at Thrissur Swaraj Round and Thiruvambadi Sri Krishna Temple at Shoranur Road. These two temples are themselves central points for approximately ten other nearby temple communities. Hundreds of thousands of people from all these communities flock to these temples to celebrate Thrissur Pooram.

It was midnight but people had begun gathering for the next day, well before the festival would officially begin. Anticipation for the big event was palpable. Vendors were setting up their shops, tents, and food stalls. A tall, makeshift temple was being decorated with strands of brightly coloured lights. Inside this ad hoc temple, a blind bull named Thriprayar Ramachandran would be shackled all night, forced to stand dangerously close to the fireworks display as people enjoy the festivities. He would remain tethered in the same spot into the wee hours, deprived of sleep. But this was just the starting point of the elephants' agony. Their suffering would be magnified a thousand-fold during Thrissur Pooram itself.

Tony and I continued our on-foot reconnaissance, capturing some footage along the way. We witnessed other elephants who were being treated in a depraved fashion. But we had to conceal our emotions and keep pushing on, shooting as much footage as time and circumstance allowed. Then we returned to our hotel around midnight so that we could get some sleep and a decent start the next day.

Just as I was falling asleep, BOOM! There was a massive explosion outside. I literally jumped out of my bed, terrified, and ran out onto the balcony to see what was happening. The obnoxious fireworks that are part and parcel of every temple ceremony had begun. They would continue incessantly for half an hour. The impact of the explosion was so strong that my bathroom window shattered into pieces, and the floor was strewn with broken glass.

It was around two o'clock in the morning when I called the reception for a hotel employee to come and clean up the mess. As he began to do so, I shot some footage of him. I couldn't have planned this, but it would make for a very significant moment in the film and shed light on the detrimental impact of firecrackers even from afar. It was impossible to go back to sleep following the earth-shattering bombardment, so I showered and did my usual morning routine. I then called Tony's room, knowing that he could not have slept through this chaos either.

It was hard to imagine how terrified the poor blind elephant must have been, shackled defencelessly so close to where the fireworks were set off. Elephants have very sensitive hearing. Their feet and trunk can feel even the most subtle of seismic vibrations. Indeed, studies have shown that they communicate through these vibrations. People can cover their ears, but this elephant would have to put up with the bombarding sounds.

As noted above, all temple ceremonies involve fireworks, of which there are several major displays. Typically, fireworks start off a day before to commemorate the grand spectacle; this, no doubt was what we'd just heard. Another major display manifests in the form of a renowned competition called *kudamattam* (umbrella exchange) between the temples of Thiruvambadi and Paramekkavu. Perhaps the grandest of all the firework extravaganzas takes place on the festival's last day, marking the end of the entire Pooram. People travel from all over the world to see it.

As I peeked out of the window, I realized that the dawn was breaking through. Thrissur Pooram is the day hundreds of thousands of people had been eagerly awaiting. After a light breakfast, Tony and I gathered our camera equipment and walked to the festival grounds. Dark clouds were hovering above our heads. As expected, there was another heavy downpour. But the rains did not dampen the spirits of the revellers who had gathered around the temple doors in the arena, awaiting the arrival of their superstar.

After about thirty minutes, a tall shadow was lurking behind the temple doors, as the familiar sounds of drums and horns became louder. In moments, a magnificent elephant pushed open the door, emerging like a rock star. He was adorned in a gold-plated caparison and his massive tusks had been polished to a glistening white. The frenzied and fanatical crowd cheered and called out his name. His regal personality could not be ignored as he stepped outside and looked around in a dignified manner, as though assessing the boisterous alien species.

At ten feet, reportedly, he is the second tallest elephant in Asia and the tallest elephant in India. Thechikkottu Kaavu Ramachandran is blind in his right eye. His body was covered in scars and his ankles draped in albino rings. These were remnants from previous wounds inflicted by the shackles that had dug into his skin. But his emotional scars were much deeper. In the past, when his suffering had become intolerable, he had unleashed his wrath, resulting in tragic endings for people and elephants. But despite his notorious reputation, the demand for this elephant continues to skyrocket. The festival mafias and scores of fans across Kerala wouldn't settle for any other elephant to usher in the Thrissur Pooram. Dragging his shackles like a prisoner, his handlers coerced this slave onto the festival's outdoor ceremonial grounds.

It began to drizzle again as the magnificent bull walked around the large arena. Thousands of people followed him with umbrellas over their heads. This elephant would have produced amazing prodigies in the wild. Yet, in captivity, he had been emasculated and rendered impotent, forced to do unnatural things like bowing before the temple deities and carrying people on his back. What a tragic waste of his dynamic genes!

Soon after this ceremony, the bull elephant was dragged away. We followed him for a short distance, filming the pitiful way he was loaded back into the truck. There was a poignant moment when the bull looked directly into Tony's camera, his intense eyes expressing silent suffering. After his truck pulled away, Tony and I went separate ways on the festival grounds so that we could get as much footage as possible with two cameras.

When we returned to another section of the temple premises, it was shocking to discover that the temple roofs had been shattered by the powerful fireworks that had been set off the previous night. Ironically, this site had been declared a protected monument by the Archaeological Society of India. However, it now looked like a target of uninterrupted artillery fire.

On the other side of the grounds was a dramatic, albeit macabre display, depicting an elephant trampling a man who had blood splattered all over his face. It seemed to celebrate the horror of the misguided myths that perpetrated fear in people's minds, which often manifested as cruelty. I had heard horrifying stories of elephant handlers getting intoxicated, thereby masking their own fears, and brutalizing the elephants to instil terror in the animals' minds. But amazingly, all of the elephants we had encountered were playful and gentle.

I suddenly spotted Jairam, the bull, whom I had met during my December 2013 visit. I vividly remember how he was being hosed down by his handlers after months of neglect and abuse during his musth cycle. During my visit, they had allowed me to hose and scrub his belly as he laid there relaxed with his eyes closed. But even his momentary pleasure turned into suffering as his handler hoarsely scrubbed his wounds and mercilessly plucked out the skin from his wounds, intentionally inflicted on his ankles and other sensitive parts during the 'breaking of spirit' ritual. Watching me hose him, several women approached the elephant. They seemed nervous at first but were still drawn to this majestic being. After several moments of hesitation, they mustered up the courage to touch him. For a country that worships these animals, most people seemed largely unaware of their gentle nature.

As I was reflecting on my experience with Jairam, my cell phone rang. It was Tony. His camera was apparently malfunctioning. The shutter had closed, and no matter what he did, he couldn't open it. The lens was fogging, and rainwater had seeped into the camera. Fortunately, Tony had captured the critical shots and I had also gathered additional footage.

Humidity and heavy rains were toxic for our cameras and created serious problems for us. We had to step away from the scene to address these issues immediately. It took us a few hours to find a tiny repair shop in an alleyway. Here, we replaced the dysfunctional camera parts and bought additional equipment to help us cope with the severe weather.

By the time we returned to the main road, it was late afternoon. The torrential rains had subsided, and the sun broke through the clouds with a vengeance. The hot and sticky weather made it unbearable to walk. How hard it must have been for this batch of fifteen elephants who were forced to stand on the road of hot tar, deprived of food and water!

Ironically, right in front of them, drunken men danced away to the drumbeats of traditional music as the hapless elephants struggled to keep their eyes open. I had heard rumours that festival elephants were drugged; they certainly looked like zombies—lethargic and overly medicated.

Then, we spotted an elephant named Guruji Padmanabhan. His hind legs were dripping blood from the shackles that had dug into his flesh. As we zoomed in on the wounds, his handler suddenly blocked Tony's camera, trying to push him aside aggressively. An altercation ensued between the two men, but Tony had managed to shoot some close-up shots of the gruesome injuries. I filmed the elephant from the other side to ensure that we obtained ample footage of this very significant scene. Clearly, these handlers knew that what they were doing was wrong. Otherwise, they wouldn't have blocked our cameras.

People most often argue that festivities are part of Indian culture, in particular, Hindu traditions. A closer look at the sacred Hindu tenets sheds light on how they are being misinterpreted and exploited. Prominent themes in Hindu beliefs include *Dharma* (duty or life's purpose), *Samsāra* (the cycle of birth, life, death, and rebirth), *Karma* (action, intent, and consequences), *Moksha* (liberation from Samsara or death), and various Yoga practices. Dharma is by far the most significant goal of a human being. Regardless of the vocation that people choose, their way of life or conducting business or carrying out their profession must embrace and express virtues such as honesty, nonviolence (*ahimsa*), patience, forbearance, self-restraint, and compassion. Pursuing the Dharma entails implementing behaviours that are in accord with '*rta*'—a principle of natural order that regulates and coordinates the operation of the universe and everything within it. It also includes duties, rights, laws, conduct, virtues, and most importantly, 'right way of living' so life on earth can flourish.

The Honourable Supreme Court of India puts it succinctly: 'Unlike other religions in the World, the Hindu religion does not claim any one Prophet, it does not worship any one God, it does not believe in any one philosophic concept, it does not follow any one act of religious rites or performances; in fact, it does not satisfy the traditional features of a religion or creed. It is a way of life and nothing more.'

Is abusing India's cultural icon 'a way of life' for elephant owners and temple authorities? Everything I had witnessed at the 'cultural festivals' was the exact antithesis of the basic Hindu tenets. Elephants are not only subjected to corporal punishment and exploited ruthlessly all the time, but they also have to suffer the loss of their families and loved ones.

People seem to be unaware or turn a blind eye to the fact that elephants are abducted from the wild at a young age, ripped apart from their families, denied their birthright to socialize with other elephants, and forced into a life of service to mankind. Just as humans commune during these celebrations, elephants also crave to bond with their own kind. So, how can any god be happy and bless those who are tormenting god's most intelligent and noble creations?

Another glaring paradox is that the elephants of India are Schedule 1 animals by the classification of the Indian government's forest department. As such, they are supposed to be afforded maximum protection. And this protection theoretically precludes their capture from the wild. It is also alleged that when they are transported from the wild, invariably, the elephants are driven through various checkpoints, money is thrown at the guards to look the other way. It's no secret that corruption is a major part of the problem. But it magnifies a thousand-fold when it pertains to elephant capture as it involves a lot of money.

Back on the festival grounds, the drunken masses were

dancing away, oblivious to the suffering of these sentient beings. The highlight of the three-day festival is the umbrella ceremony which is referred to as *kudamattam* in the native dialect of Malayalam. The ceremony is a contest in which two rows of bull elephants, representing the two rival temples, are lined up at opposite ends of the fairgrounds. Young men are then hoisted onto the elephants' back and stand on their delicate spine, raising umbrellas to the skies. Even priests, whom I interviewed, struggle to reference any Hindu scripture that supports this practice.

During one of the performances at this event, a twenty-two-year-old elephant named Adiyattu Ayyappan was spooked by the sounds of pistols that were used to shoot confetti. He turned around distressed, ready to break loose and bolt. However, using cruel tactics, his mahouts brought him under control. Ayyappan's nervous owner then quickly appeared at his side, trying to calm him down. Often, such provocative actions push the elephants to run amok, kill people, and destroy properties. Eventually, they are captured using spiked chains and then tortured for days on end for 'misbehaving', whereas in reality, the youth are to be blamed.

The next day, during my interview with Ayyappan's owner, I asked him about the ignorant behaviour of those who had been shooting the confetti. I also grilled him about the way the bull had reacted, which could have been disastrous. The owner gave me a vague answer. Apparently, even he had been caught off-guard by the actions of the revellers in shooting off the confetti that had spooked Ayyappan. Having owned elephants for several years, his response makes you wonder whether he knew anything at all about the sensitive nature of this giant.

We later heard that four elephants ran amok at the Thrissur Pooram in 2014. A young bull, who was made to stand on a hot concrete floor, broke his chains and ran from the temple, out of control. Apparently, his sensitive feet were unable to handle the blistering heat of the tarmac. He gave his pursuers the run-around for thirty minutes and, eventually, they had to tranquillize him.

The sea of people, the chaos and confusion, the polluted air filled with chemicals from the fireworks, and the sizzling heat were all too overwhelming. But regardless, every person we spoke with justified the exploitation of elephants by twisting the meaning of the holy scriptures. Given that elephants are considered the embodiment of the Hindu god, Ganesh, and are believed to bring good luck, they are displayed in all kinds of ceremonies, store openings, weddings, and even political rallies. But people don't even stop for a moment to think, how exploiting and torturing any of God's creations would bring any luck at all? By not delving into the deeper meaning, they were betraying the very Hindu tenets they claimed to respect. Their ignorance was appalling. All in all, this was the worst kind of elephant torture I had ever witnessed.

Over the next ten days, Tony and I interviewed at least fifteen people in Kerala. They included an elephant owner, Lakshmi's handler, veterinarians, and, for a second time, the popular activist, Venkitachalam. We also interviewed a veterinarian who was employed by the government. He had a notorious reputation for issuing fake fitness certificates so even ailing elephants can be paraded in festivals. He certainly didn't expect a barrage of questions from me. Indeed, when he was unable to respond, he dismissed my question by stating it was 'stupid'. I also contacted the man who had been the chief wildlife warden in 2014. I wanted to get his overall perspective on elephant management and welfare in Kerala. However, he declined our interview request.

What struck me the most in all these dialogues was the cognitive dissonance. People had conflicting beliefs. The owners and temples tortured the elephants, who they professed to love—revering and defiling them at the same time. It was disappointing that my native home of Kerala had become the ground zero of elephant torture that had reduced India's cultural symbol to commodities for profit.

In the ensuing days, the festival organizer, Menon, arranged a trip for us wherein we could film the stunning backwaters of Kerala. He joined us, together with his teenage grandchildren. During our time together, I had the unique opportunity to get into the minds of the youth who were part of elephant clubs that promoted the use of elephants in festivals. However, when questioned, the teenagers vehemently denied that elephants were being abused. They pointed out the way dogs and cats are leashed in North America. They were so set in their thinking that they fervently rejected my suggestion that the sheer size of elephants makes it difficult to control them without utilizing tremendous force and torture. They also cited the ruthless treatment of cattle in the developed world. Bringing this heated debate to a close, I argued that two wrongs don't make a right.

Clearly, my world views had changed significantly. Although my cultural upbringing in India had fostered an attitude of superiority over animals and other living beings, my thinking had evolved over time. This was due to the education I'd had, the travels I'd been fortunate enough to pursue, and the awareness I had developed as a result. Perhaps after migrating to Canada more than three decades ago, I had learnt to view animals differently.

I placed myself in the teenagers' shoes and realized that when I was of their age, I used to be like them—because I hadn't known any better. They reminded me of how I, too, had gone with the flow when I lived in India and how much of a conformist I had been. How then could I blame these people when I myself used to be like them not so long ago? It became clear to me that without enlightening people with knowledge, the darkness of ignorance will linger. Educating people about the elephants' plight and the abuse they suffered at the hands of humankind is necessary to spur conscious actions that would improve the lives of elephants.

The number of festivals in Kerala keeps rising by the year, and so do the number of elephants exploited for profit, resulting in a significant spike in the number of elephant deaths. More religious institutions are emerging across India and bringing elephants to their institutions behind the veil of culture and religion. Most of the elephants are bought and transferred from India's northern states of Bihar and Assam.

In my estimation, 'more' is the common thread that is being woven into the cultural tapestry. These desires to want more are unending. No matter how many elephants are paraded, no matter how much profits are made, no matter how many temples are built, people will never be satisfied. Is there any end to it?

Everywhere I looked, all I could see was suffering, torture, and unconscious behaviour. People were following the processions blindly and revelling obliviously, completely indifferent to what the animals were going through. The youth were simply standing and watching as mute and passive spectators, dancing and wasting away their lives. They seem to have lost their sense of purpose or service.

What it boils down to is this: Humans can uphold their cultural values without torturing elephants, or any animal for that matter. There are other ways to celebrate faith and culture. It's certainly not necessary for animals to be used in ceremonies. Thankfully, traditions without elephants do exist in India. Our newly found friend Menon was able to give us a taste of them as well. For instance, Kathakali, which originated in Kerala, is a creative blend of music, dance, and acting that dramatizes epic Indian stories such as Ramayana and Mahabharata. The performers wear elaborate costumes and dramatic make-up, their faces painted to perfection.

We also went to a village where the native people performed snake dance in a ritual called *Sarpam Thullal*, which paid homage to the cobra. In India, snakes, especially cobras, are mentioned throughout the Vedic books. But in this ritual, no cobras were used. Instead, an intricate recreation of a cobra had been created with powdered grains and spices on the temple floor.

In addition to capturing footage of the elephants, we also gathered videos of Kerala's natural and cultural beauty in order to present a balanced and objective perspective of this tiny state located at the southern tip of India. This footage also allowed us to highlight the authentic nature of the natives. They opened their generous hearts and homes and afforded us, strangers, delicious meals, no matter the time of day or night. These were some of the most touching and confusing moments of my journey. Most often I struggled to understand how such kind people, the very same people who had attended Thrissur Pooram, could not see the pain and suffering of the elephants.

Our time in Kerala was nearing an end. But the day before our departure, Tony and I were invited to attend a bizarre ceremony, as though we were preordained to witness the most unambiguous paradox. In this ceremony, an elephant named Chandrashekaran, who had died in 2002, was commemorated. Fifteen enslaved elephants were lined up on the same festival grounds where Thrissur Pooram had taken place. Among them was a stunning-looking elephant named Ramabadhran. As with all elephants, he was shackled severely, his eyes displaying sheer hopelessness. He had large patches of depigmented skin all over his body and albino rings on his ankles. As I observed him keenly, I noticed that he was struggling to eat. His trunk was paralyzed.

Different stories swirled around on how this happened. One theory was that after Ramabadhran was loaded onto a truck for the festival circuit, the driver inadvertently slammed the door on his trunk. Another story stated that he was bitten by an insect, which caused the paralysis. Yet another theory was, he was beaten so severely on his trunk that he became paralyzed. Regardless, it was distressing to watch this poor animal suffer in this manner. As though this was not torturous enough, his handler dragged him to a nearby water tank where he tried to scoop up water from the tank with his lifeless trunk. He dipped his trunk in the water a few times, then lifted it up only halfway where it dangled. It was as if he was displaying his pathetic condition for our cameras.

Can you imagine being thirsty but unable to drink water, especially when you are holding it in your hand? After about fifteen minutes of this sad display, his mahout barked out a command. At this point, Ramabadhran's trunk flopped lifelessly out of the tank and he hobbled feebly away. Despite his pitiful condition, the handler refused to loosen his shackles. In fact, he poked the poor animal's leg with a bull hook. He also yelled at him, yanking at the shackles of the meek being who so silently obeyed his orders.

Suddenly, I had a eureka moment. It dawned on me that I, too, was shackled, paralyzed, and crippled by fear-based thoughts that my mind had created. I also understood that this emotional handicap was working to prevent me from expressing my authentic gifts. I had become paralyzed by self-doubt inflicted from my past conditioning and shackled by the thoughts and beliefs of others that had been imposed on me when I was a young girl.

Although I was now in India on a mission that I deemed to be sacred, I still dreaded going out on a limb for the elephants, fearing that I would be ridiculed if I failed. In just the same way, Ramabadhran had been conditioned by terror. This elephant made me realize that I was shackled by my past. Its burden of guilt, shame, and resentment weighed me down and impeded my ability to be fully present in my life and in my work.

My father may have punished me when I was a young girl. But decades later, and even after my father's death in 2012, I was still clinging to the painful events that had happened so many years ago. Although on his deathbed, my father had shared his own suffering as an orphaned child, I had still been unable to move forward. I thought that I had mended my splintered relationship with my parents, but Ramabadhran shone a light on the traces of resentment that still needed to be dealt with.

In the last three years of his life, Ramabadhran had developed serious foot infections and massive tumours on his right hip. These became severely septic, further deteriorating his health. But this elephant's ill health did not prevent the temple authorities from parading him on the temple's granite floor on 17 March 2017. The vets had given him a clean bill of health, stating that he was receiving proper medical treatment. That fateful day, he ran amok, unable to bear his agony. The vets had been terribly wrong in their assessment. Or perhaps they were manipulated by the elephant establishment to give him a fake fitness certificate, which is something that happens only too frequently. Just over a month later, on 26 April 2017, he died. He had been shackled all his life and was shackled when he died at his tethering site in Thrissur. Ramabadhran's death was inevitable. It is a miracle that he lived and endured so much suffering for so long.

He was unable to roam freely in the wild when he was alive. But, at least, in death, Ramabadhran could reclaim his basic birthright—freedom. His tormented body is no longer on this planet, and his spirit will never again be paralyzed. His captors sang his glory in his death after torturing him when he was alive. This is the ultimate absurdity!

Ramabadhran's story is prominently featured in *Gods in Shackles*. He was among the fifteen elephants who had been starved all morning and paraded beneath the scorching sun to honour an elephant who had died fifteen years prior due to torture. We followed him closely, and I had a chance to interact with him when we filmed. One of the greatest memories I will always cherish is feeding him cucumbers, watermelons, and bananas after the commemorative ceremony. Those few moments of bliss will remain etched in my mind until I, too, leave this mortal coil. Through his suffering, Ramabadhran shone a light on my dark emotions and helped me realize that fear is paralytic. Just as his paralyzed trunk was lifeless, fear had paralyzed me and made me feel lifeless.

Back at that ridiculous ceremony, hundreds of people had gathered to commemorate elephant Chandrasekaran's death. I suddenly heard my name being announced on the microphone. The president of the temple was introducing me to the crowd and asked me to stand up to be acknowledged. He said I was originally from Palakkad, Kerala, and now lived in Canada. With a slight tone of humour in his voice, he also announced that I loved elephants but hated to see them tortured. The two veterinarians whom I had interviewed earlier in the week were present in the audience. They surrounded me as the onlookers cheered on. It was all too much for a humble film-maker who was trying to produce a documentary under the radar.

As I was addressing the crowd, Tony snuck up from behind and tapped me on my shoulder, pointing to a star elephant, Shiva Sundar. This elephant was just then entering the complex, bearing the plaque of Chandrasekaran who was being commemorated. I picked up my camera and resumed filming as the handler atop this elephant was ducking his head trying to get through the low entrance.

Right behind him was the only female elephant among the fourteen bulls—my beloved Lakshmi. I moved beside her and as I did, she stretched her trunk and caressed my palm, sniffing affectionately. She was the second elephant to be taken before the altar. Here she was jabbed with a pole until she saluted with her trunk in a trained behaviour to the portrait of the dead elephant that had been placed before a massive burning lamp. Then she was given some fruits. The other elephants followed her.

Somehow, I felt these elephants knew what they were doing. Numerous studies have shown that elephants are self-aware and capable of recognizing objects before them. I suspect, they knew that the portrait belonged to one of their own.

After the ceremony ended, one of the temple's trustees

and C. A. Menon's son, Narayanan, shared the sad tale of how the elephant Chandrasekaran, whose life they were commemorating, had given up his last breath. Apparently, during the final days of his illness, he became restless and wanted to be left alone. But he would welcome Narayanan and calm down in his presence.

On the final day of the bull's departure from the earth, several people had gathered around him. Suddenly, a bearded man dressed in a saffron robe appeared. Apparently, nobody had ever seen him before. This saint told the crowd that unless the elephant was released from the shackles he would not die. Narayanan then demanded that the chains be removed, but the timid handler refused, fearing that Chandrasekaran might 'spitefully' attack him. However, Narayanan went against all odds and unshackled the elephant who, just five minutes after he was freed, flung his leg freely up in the air, and took his final breath.

It was hard not to shed a tear or two, listening to this sad saga. I told him how grateful I was to him for giving this elephant dignity and the basic right of freedom before he died. As a gesture of honouring this elephant, they were planning to erect his statue in front of the temple gates. Here again, I observed the hypocrisy and dissonance. Elephants are tortured to death and revered after they die. Processions are organized wherein elephants are tortured to mourn the death of an elephant who had also been tortured—to death. Commemoration in public and abuse behind closed doors.

People seem to be chasing after experiences that offer them fleeting happiness and instant gratification. Eternal joy does not require the torturing of elephants for entertainment. In a highly advanced technocratic era, there are many other amusements available for exploration. Almost all the young men at the festivals carried iPads, notebooks, iPhones, and other sophisticated gadgets that allowed them to take selfies and make videos. Although technologically evolved, most of them seemed consciously unevolved.

But this situation is not unique to India. Around the world, as humanity continues to chase after technological and economic advancements, the integrity and well-being of many different species of wildlife—and, indeed, of nature itself—are being compromised. People are so caught up in pursuing the false gods of materialism that only the self-aware ones bother to look within. Nor are humans interested in evolving consciously, which would include a developed sense of compassion for all living things.

In any event, our time in Kerala was nearing the end and so was my time with Lakshmi as we had one last shoot at 4:00 a.m. on the day we were leaving. At this point, we would film her being bathed in a stagnant tank. Her wake-up call, which was so early in the morning, was just as rude as mine had been growing up.

Every day, during her early morning prayers, my mother intentionally, for reasons known only to herself, rung the ritual bells at 4:30 a.m., when the world was fast asleep. After the rude wake-up call, I would drag myself out of my bed, just like Lakshmi did every morning.

The morning we were there to film her was no exception. After her handler had pushed and kicked her a few times, Lakshmi rolled her huge body over and stood up. As she dragged her sleepy self towards the tank, she stopped for a moment, implying that she wanted to dump. But the impatient handler intimidated her into getting into the tank. It was filled with contaminated water. After a few moments, she defecated in it. He picked it up and tossed it outside the tank and went on to soak her in the same contaminated water. At least she was hosed down with clean water as she exited the tank. She was then fed a scoop of leftover rice before ambling off to the temple to perform her morning ritual.

During our time in Kerala, I ran into Lakshmi almost daily. This occurred through planned and unplanned encounters, as though we were meant to see each other frequently. It was now hard to say goodbye to my beautiful soulmate. But it was only a matter of a few months before I would return to Kerala, at which point I would see her again.

Looking back on the dissonance, the love-hate relationship between the elephants and the handlers, the ignorance, the denial, the justifications, and my own judgements, I needed to come to a place of acceptance and understanding. The schizophrenic nature of this love-hate scenario was a conundrum that only enlightenment and evolution could solve. Ignorance, illiteracy, and cultural myths have blinded the festival attendees, temple authorities, and elephant handlers and owners so much that they seemed unaware of the emotional and physical agony that these innocent animals endure day in and day out.

It seemed to me that I had been put to face the greatest test of my life—to love and accept people who acted in unloving ways even towards the most loveable animals. I had to be more tolerant and patient, just like the elephants. I could only hope that as I did the work on myself, something inside these other people would begin to shift as well.

With a heavy heart, I left Lakshmi behind, in addition to all of the other abused elephants of Kerala. However, my resolve to expose their suffering and bring an end to it had strengthened. As much as my heart was shattered, I had no time to brood over my emotional pain. Tony and I were on the move, travelling to our next shooting location. Fortunately, the worst was over. We had finished filming the most gruesome scenes—at least for a short while—and we were ready for an adventurous journey into the jungles of India.

Chapter 9

My Wild Adventures

Tony and I arrived in Bangalore in the wee hours of 14 May 2014. Despite having had only a few hours of sleep, by 6:00 a.m., we were energized and ready for our adventurous journey into the jungles of Karnataka. Varma and Dr Raman Sukumar, whom I had met during my December 2013 fact-finding mission, had arranged this special journey into the forests near Nagarhole. Here, wild elephants convened at this time of the year. My intention was to portray the stark contrasts between wild and captive elephants in the hopes that people would be empowered with knowledge, which, in turn, could propel attitude change and an outright ban of elephant captivity in religious institutions, circuses, and zoos.

The scientists warned us that there had been only a few sightings of elephants recently as rainfall had been scanty. But we took our chances. After a good three-hour drive, we arrived at the forest entrance. Two female guards laboriously opened the entrance's heavy wooden gates, and the driver revved up our jeep. Entering into the uneven terrains of the jungle, our jeep stirred up a dense plume of red mud. Shortly thereafter, several baboons blocked our unpaved pathway as though reminding us that this was their territory. We waited for a few minutes, but they refused to move. Suddenly, the trees ruffled and a large herd of spotted deer—approximately fifty or so—dashed into the bushes. Perhaps they had been threatened by a predator. The baboons scrambled and suddenly disappeared, clearing the path for us. Our driver, meanwhile, warned us that a large carnivore could be lurking in the woods and asked us to wind up the windows. The suspense was at its peak as all eyes scanned around silently to spot a tiger or leopard, but no such luck.

As we got deeper into the jungle, flame trees with bright red flowers and the sun's rays dancing through the canopies cast unique shadows on the forest floor. Bison, impala, peacocks, and rare birds congregated near water holes, and the scent of wildflowers filled the humid air. For the first time on this entire trip, I could breathe deeply and feel the rhythm of my heart in this lost paradise. Just then my gaze suddenly fell upon my watch and I realized that two hours had gone by inside the forest. But we hadn't spotted a single elephant! I turned to Tony who was busy filming. Varma, however, noting the disappointment on my face, shrugged his shoulders, reminding me that there were no guarantees.

In that very moment, I noticed a herd of large grey animals far away and pointed them out to the driver. He drove as fast as he could on the forest trails. But by the time we got to where I thought I had seen the creatures, they had disappeared. We waited patiently for a while. Lo and behold, almost thirty minutes later, a large herd of elephants emerged. They were protecting a baby. It was indeed a breathtaking scene to witness so many elephants—at least, a dozen of them—moving towards a meadow to graze. As they fanned out into the open space, we spotted two adorable baby calves, constantly touching their mother with their tiny trunks, seeking comfort and some reassurance. The protective relatives shielded them from the aliens in the jeep, just as humans would do with strangers.

After witnessing the tormented elephants of Kerala, I treasured every moment watching these mesmerizing animals in the wild, living out their true nature. They were doing exactly what they were meant to do—socialize, bond, protect their babies, graze, and communicate in their own language. I wondered what it would be like to be part of the herd and become one of them. I treasured every moment of being in their serene and peaceful world. These magnificent animals were carefree and chain-free—none of them were blind, there were no shackles, no wounds on their legs, no shortage of food or water for them to consume. Here, in the wild, they could eat and drink as much as they wanted, whenever they wanted. I only wished that we humans could live like elephants, coexisting harmoniously with other living beings.

In a few moments, another herd joined the first one. This second herd began to graze just yards away from our jeep. Overall, there were at least two dozen elephants, most of them female. Our cameras loved every picturesque frame of these magnificent animals who were so beautifully offset by a stunning backdrop of majestic mountains.

After gathering enough shots from unique angles, we subtly reversed the jeep and veered off on a different path. Just five minutes into the drive, we saw a magnificent bull elephant with one tusk digging it into a tree to rip off a large piece of bark. He then kicked the earth beneath his feet. Extracting roots from the soil, he scooped them up with his trunk, dusted them off on his legs, and tossed the delicious meal into his mouth. The two scientists who were with us speculated that one of his tusks must have fallen off, or perhaps he had been born with one tusk. We then spotted another bull without any tusks. They called him *makhana*. He was standing in the marshy area, content in his sweet world. He scooped up slushy mud with his trunk and tossed it onto his back, trying to cool off from the merciless heat.

Later, we met a small herd of three elephants with a baby, just weeks old. He strayed a bit too far from his mom, so his protective aunt stepped into the picture to nudge him back in between the adults. It's incredible how much humans have in common with these amazing animals. They socialize and protect their young ones as we do. None of them were hostile toward us. Instead, they went about their business despite the fact that we had invaded their space.

As I was treasuring this dream with the wild elephants, our driver reminded us that the sun was beginning to set. It would take us an hour to get out of the jungle. As difficult as it was to leave, my first few encounters with wild elephants roaming freely in the Indian jungles created an indelible impression in my mind. Tony and I returned to our hotel after an easy day's work. We retired early, for more adventures awaited us in the morning.

Before the dawn broke, we rode a boat up the Kabini River and encountered a large herd of elephants on its banks. As our boat approached them, the family huddled together, trying to protect a calf. The matriarch then launched a group discussion. Using tactile language, they touched each other with their trunks, trying to figure out how to respond to us. After a few minutes, the herd walked away, with the exception of the aunt who came charging toward us. She was trumpeting furiously, demanding that we back off. Our captain had no choice but to turn around before we passengers got into trouble. On the other side of the river, more elephants were heading towards the water to drink and bathe.

Later in the day, Varma and I took a bus trip into a different jungle, kind of a safari, run by the state government. As soon as we entered into a dense section of the forest, a bull elephant who was covered in slush became so agitated that he came dashing towards us. He could not bear the repugnant sound of our vehicle and the people in it. Fortunately, the driver reacted quickly and sped away. It could have been disastrous.

Further along the way, we spotted at least five herds of elephants, accompanied by sub-adults. Totally unfettered by our presence, they continued to graze and socialize. That afternoon was truly magical. We shot footage of every single elephant we saw so that we could offset these delightful visuals with the horrific ones, portraying the tormented elephants of Thrissur Pooram and other Kerala festivals. Although it was hard to get decent footage from the crammed bus, with too many people, we still had an adventurous time.

The wild elephants in their elements seemed reflective and deliberate. They honoured each other's space and their defined roles within the herd, allowing them to express their innate nature.

It was hard to comprehend the insurmountable emotional and psychological torture that captive elephants must suffer. Unable to toss mud on their backs or even scratch their own bodies, they are deprived of basic freedoms and stripped of their dignity. They are reduced to slaves to serve the whims of their masters. Shackled day and night, they live a desolate life and die a miserable death. It was difficult to understand how conscious human beings could capture and decimate these awesome animals, pushing them to the brink of extinction in the process.

Does this sound familiar? It had taken bloodied riots, civil rights movements, and someone as courageous as Abraham Lincoln to take bold actions against human slavery. But even in the 21st century, injustices prevail. For instance, human civilization is still trying to deal with inhumane issues such as child trafficking, child labour, child prostitution, subjugation of women, and racism, among other things. The International Labour Organization estimates that even today, over 40 million people are in some form of slavery. Perhaps it would take decades of rioting, animal rights movements, and a powerful leader like Lincoln to ban elephant slavery.

After three heavenly days in the wild, it was tough returning to reality. On our drive back to the hotel, Varma lauded a woman by the name of Suparna Ganguly. She is the founder of Wildlife Rescue and Rehabilitation Centre in Bangalore City, which is considered to be the IT hub of India. She had collaborated with Dr Sukumar and the Animal Welfare Board of India (AWBI) in order to conduct an extensive investigation on the captive elephants of

Guruvayur Temple, a United Nations heritage site. In this, she had been instrumental in releasing a groundbreaking report that corroborated the footage I had gathered at the Guruvayur Captive Elephant Sanctuary. Intrigued by the wealth of information she could provide, I promptly phoned her to arrange an interview. Fortunately, I was able to reach her. Typically, she's on the move a good part of the year. But as fate would have it, she was not travelling during the next week, and I quickly made an appointment to meet her the following day.

Equipped with our camera gear, Tony and I arrived at Suparna's place the next morning. A woman with a crown of salt-and-pepper hair greeted us with a warm smile at the entrance to a luxury apartment building. Suparna was wearing a white cotton kurta (long shirt) with flowers printed in aqua blue and a turquoise blue shawl wrapped around her neck to match her traditional Indian pants. She led us into her office and offered us a cup of hot chai. After a brief chat, she graciously invited us to join her family for lunch. We were then led through some spiral wooden stairs to a large terrace sprinkled with potted plants. Here, cane chairs with comfortable cushions had been arranged for our interview. As Tony was setting up his equipment, Suparna and I became deeply engrossed in a discussion about the overarching philosophical and ethical challenges related to animal welfare issues in India.

Almost twenty minutes had passed, and Tony was still struggling to get the camera working. We were having the same issues with the intense heat and humidity that had hounded us all through our filming in Kerala. Suparna's story was too important to be missed and we were desperately trying to revive the camera for this last interview. However, we knew that only time would recalibrate the device. The jovial atmosphere suddenly soured and became tense. But fortunately, almost forty minutes later, the moisture evaporated. The camera rolled, as Suparna poured out her heart and soul during a two-hour-long interview. She clearly articulated the nexus of culture, commerce, and corruption.

Although capturing of elephants had been banned in India since the year 1982, it still continues unabated to this day. It begins with the abduction of a baby or sub-adult male elephant from the northern forests of Assam, Bihar, or Arunachal Pradesh and transporting the animal through checkpoints by bribing the local forest guards. These innocent animals are then subjected to brutal training for a week to break their spirit. The training entails deprivation of food, water, and sleep, in addition to chaining them with spiked chains and brutal beatings with vicious weapons like bull hooks. This 'ritual' takes place for, at least, five relentless days, with different handlers taking turns, until the bull surrenders. He eventually realizes that he has lost freedom, and utterly exhausted, he begins to 'obey their commands' just to get some food and water. He is then transported illegally through interstate boundaries and brought to Kerala, where the owners are given fake certificates by corrupt government officials. Elephants then have to undergo more brutal training until the animals become acculturated to the local handlers' commands and language.

But people seem to be totally unaware—or are wilfully blind—of these unimaginable realities and horrible practices that the so-called festival elephants have to endure in order to entertain them. It's hard to imagine the trauma that the poor bull would have had to go through, beginning from missing his bachelor friends, his wild home where he could roam freely for as long as he wanted, and find mates when he needed.

Greed, selfishness, corruption, apathy, and complacency are part of the deeper issues decimating our planet's natural world and its inhabitants. Elephants, for instance, arrived on the planet some 80 million years ago, whereas humans arrived just a few thousand years ago. According to a study entitled *Biomass Distribution of Earth*, humans make up only

0.02 per cent of the earth's biomass. Yet we have destroyed 83 per cent of the wildlife due to our reckless actions.

Religious institutions are supposed to set a shining example by conducting themselves ethically and by abiding the laws of the land. Instead, they flout the government regulations blatantly, hiding behind the curtains of religion and culture. They exploit the vulnerabilities of politicians, who would dare not stoke the cultural and religious sentiments, lest they might lose the votes of hundreds of millions of devotees. If only elephants could vote . . .

Corruption in these religious institutions is rampant. The chairman of the board of directors at these temples are political appointees who also own or at some point owned elephants, which makes it difficult, if not impossible, for an honest chief wildlife warden (CWW) to carry out his duties. Everything becomes politicized, with elephants and wildlife caught in the crossfire.

Meanwhile, Suparna explained that enormous amounts of cash are also exchanged to ensure that no paper trail is left behind. A 'handsome bull elephant' can cost up to one crore in rupees, almost $US145,000. The owner not only has to recoup all that money, but also generate profit. In concert with the temples, the owners begin to concoct festivals so that elephants can be paraded beneath the scorching sun. They're transported in trucks, chained precariously, driven through narrow and bumpy lanes, making it impossible to even balance their bodies. They're sometimes driven non-stop for eight to ten hours, especially when they're transported from different districts.

It was remarkable how, with her interview, everything began to fall into place. Our journey had begun with the Thrissur Pooram festival, which revealed the dark truths behind the glamorous festivals and the abhorrent treatment of elephants. Our trip ended with Suparna's deep insights, as discussed above, which helped connect the dots between the nexus of culture, commerce, and corruption. A larger force was directing, guiding, and informing the work I had undertaken, with my earnest intentions. My heart was overflowing with a profound sense of gratitude for a smooth unfurling of my path at that point.

After a day filled with warmth and enlightenment, it was now time to leave. Just as Tony and I were stepping out of Suparna's apartment, she casually mentioned that I should connect with a poet in Kerala by the name of Sugathakumari, whose name I quickly jotted down.

Suparna and I stayed in touch throughout my post-production in Toronto. Her passion and determination to be a voice for the voiceless elephants was truly inspiring. Aside from our cultural connections, we also shared similar values and world views. We had organically cultivated mutual trust, respect, and acceptance on a profound level. In the days ahead, we turned to each other for moral and emotional support. But I never imagined that, in the ensuing months, Suparna would be key to helping me attain a significant milestone!

That evening, as soon as we returned to our hotel, Tony became sick with nausea and fever. His body could no longer handle the severe heat, pollution, and spicy food. But there was still one final mission scheduled for the next day. We were due to visit the Bannerghatta Biological Park (BBP) where an elephant named Sundar was expected to be released after his rescue from a notorious temple in Kolhapur in Maharashtra state. However, given Tony's ill health, I was left to my own devices, literally.

The following day, equipped with my camera, I was dropped off at the park's entrance. It was packed with tourists eager to venture into the forests. As the first jeep arrived, a crush of people scrambled to get into them. Fortunately, due to prior arrangements, I could sit in the jeep's passenger seat, right next to the driver, affording me a great vantage point from which to film.

The BBP is 260 square miles and owned by the government of Karnataka state. In it, deer, panthers, and other creatures roam freely. Hundreds of people are employed on its grounds, with hundreds more in the office. Inside, the Central Zoo Authority (CZA), a central governmental body, had set aside a true elephant sanctuary where sixteen rescued captive elephants receive the care and respect they deserve. These elephants were between the ages of six months and sixty years old. The adults were rescued from temples and circuses across southern India. They produced babies after mating with wild elephants that frequented the park from the hills beyond.

There was a large open space that featured a deep pool. On the banks of it, four female elephants waited patiently for a bull elephant to get out of the water. But he was having too much fun splashing it all over his body. Apparently, this handsome bull, Vanaraj, had also fathered a baby. I later learnt that he was an abused temple elephant from Kerala who had been rescued by one of the animal welfare groups. His life had changed forever after he was released into this semi-wild sanctuary. Now, he gets to do whatever he wants—swim in the pond, drink water when he is thirsty, eat when he is hungry, and socialize with the ladies. After having been traumatized, these elephants can experience at least some semblance of freedom here.

It was heartening to learn that the rescued temple elephant, Sundar, would be in good company in this new home. But I was still unsure how or whether I would even meet this elephant, let alone how I might include his story in the film. Regardless, I gathered some footage of the park. The next day, 22 May 2014, Tony and I flew back to Mumbai. He returned to the United States, while I decided to spend some time with my family. Or, at least, that's what I had planned. But the universe had other plans for me, for more synchronicities awaited me the next day.

Chapter 10

The Sorry Saga of Sundar

In an earlier chapter of this book, I had mentioned the crowdfunding campaign that I had launched to raise money for my film production. At that time, a Facebook friend, Sujata, posted news of my campaign on a page dedicated to exposing the sad plight of Sundar, a temple elephant, in a place called Kolhapur, not too far from Mumbai. This is how I had initially learnt about Sundar's miserable life.

This magnificent bull elephant had been taken from his family when he was eight years old, tortured into submission, and sold to a politician in a small village in Kolhapur district. This politician had 'donated' this sentient being to a religious institution called Jyotiba Temple. For six tormented years, the poor elephant was chained beneath the scorching sun on the temple's hard granite floor, forced to 'bless the devotees' by placing his trunk on their head for a donation to the establishment. Sundar was the poster child for all the suffering temple elephants of India.

After receiving numerous complaints from devotees, the world-renowned animal welfare organization, People for the Ethical Treatment of Animals (PETA), intervened. I managed to link up with PETA's co-ordinator in India, and he set up an interview with their veterinarian in Mumbai, who was knowledgeable on the issues at hand. He explained that PETA had launched an investigation into this elephant and its keepers. Using an undercover camera, they had captured Sundar's abhorrent treatment—his handlers beating him mercilessly as the poor elephant cried in agony. His right ear had been ripped with the vicious bull hook. Some skin beneath his eye had also been torn by the same weapon, nearly blinding him.

The video I had seen about Sundar had gone viral on social media. Sundar's plight made angry waves around the world, invoking popular celebrities like Sir Paul McCartney to speak out. He wrote a letter to the prime minister of India saying, 'I have seen photographs of young Sundar, the elephant kept alone in a shed at Jyotiba Temple and put in chains with spikes. I appeal to you to do what is right here and get Sundar post-haste to rehabilitation in the forest. Years of his life have been ruined by keeping him and abusing him in this way and enough is enough. I most respectfully call on you to use your authority to get Sundar out. The whole world is watching.'

PETA asked the state forest authorities to confiscate the animal. Their repeated requests were ignored and Sundar continued to suffer. Animal activists launched a passionate global campaign demanding Sundar's release. But nothing happened. Subsequently, PETA launched a case at the High Court of Bombay, which issued orders to release Sundar in a sanctuary. For over a year, the bureaucracy ignored the court orders and dragged its feet. Meanwhile, the politician who had donated Sundar to the temple used his influence to fight PETA every step of the way. Sundar was already enslaved in chains and now, he was also wrapped in red tape.

As genuine scientists and animal activists in India were trying to rescue this abused elephant, global citizens began to criticize the same people who were trying to help the elephant, assuming that nothing was being done. Even as gradual changes were happening on the ground, the activists were demanding immediate relocation of the elephant, without understanding the complexities of a corrupt system.

After having lived in Canada most of my life, my own assumptions about people who worked on the animal welfare side of things in India were bleak. But after gaining a better understanding of the political system, I realized that cutting through the bureaucratic processes required equally smart tactics as the ones implemented in saving animals. I was convinced that everyone was doing their best to rescue Sundar. Despite being one of the largest democracies in the world, freedom of speech and of the press in India, has been suppressed by the powerful and wealthy who employ tactics of intimidation to obtain results that suit their personal agendas.

After I finished the interview with the vet, it became clear that Sundar was still trapped in the hands of his abusers. On my way back, I kept wondering what more I could do to help this suffering elephant. The next thing I knew, my brother offered to lend me his chauffeur and arranged for him to escort me wherever I needed to go.

Now, get this. Sundar was in the same village that this driver was originally from. But it gets even better! The infamous Jyotiba Temple, where Sundar was housed, turned out to be the driver's own family temple! Nobody other than a higher power could have aligned these synchronicities.

Every step of the way, I was receiving clear evidence that my journey was being divinely orchestrated. I had to constantly remind myself to take a step back and acknowledge that I was merely being used as an instrument to serve the spirit's mission. All I had to do was have good intentions, surrender my ego, and show up to do what I was being guided to do. With a bounce in my step, I announced with conviction that I would indeed meet Sundar and, in doing so, I would be escorted by my brother's driver.

Holding onto high hopes, I left Mumbai with my brother's driver the next morning at 4:00 a.m. to find Sundar. It was a ten-hour journey through rugged roads and rocky hills. The highway stretch from Mumbai to Pune was smooth, but the balance of the journey was bumpy. We also stopped occasionally to take in the breathtaking views of the lush valleys and mystical mountains from higher altitudes. We reached Kolhapur at around 4:00 p.m. and then drove another twenty kilometres to the infamous Jyotiba Temple. From a distance, its spectacular gold-plated tomb glistened on a sunny day and the temple's signature orange flag on the shiny roof fluttered in the wind.

After a steep drive, we arrived at the entrance of the temple. Here, hundreds of vendors were selling garlands, coconuts, and fruit as offerings to the deity. I spotted a snake charmer playing a flute, while a timid cobra coiled in a basket raised its head and swayed tiredly. The cobra's poisonous fangs had been extracted, and its mouth was sewed with a thin wire thread.

Snakes have a long association with the Hindu faith, indeed with Lord Shiva himself. This deity is frequently depicted with a snake around his neck, representing Shiva's powers of destruction and recreation and his ability to orchestrate these forces at will. Cobras also have a special place in the Vedic books. In Indian mythology, the snake primarily represents rebirth, death, and mortality, depicted by the casting of its skin and being symbolically 'reborn'. In many parts of India, cobras or *nagas* are engraved in stones and displayed in shrines where people offer food and flowers, and light lamps. A cobra, which is accidentally killed, is cremated like a human being; nobody would kill this snake intentionally.

But here again, this abusive treatment did not reflect the tenets of those Vedic books. As seen with elephants, snakes were worshipped and defiled at the same time. The paradoxes were all too jarring.

We walked into the temple after removing our sandals, which is customary in India before entering a holy place. The Jyotiba Temple is known for its pink theme. Bright fuchsia-pink powder covered the stone floor and people sported pink *bindi* on their foreheads. Pilgrims nearby were quietly praying. Despite the dark stories I had heard about the Jyotiba Temple, there was a semblance of serenity within its walls.

As I circled the inside of the building, I spotted a priest seated in lotus position beneath a banyan tree. I asked my driver and escort in Marathi—the native language—to inquire about Sundar. The priest told us that the elephant had left the temple six months back and directed us to a dump where Sundar used to be tethered at night. Looking at the filthy site was gut-wrenching. Construction rubble was strewn all over the place and the stench was unbearable. This majestic elephant was forced to stand on the uneven floor and sleep in that horrible place, inhaling its nasty stink. What a hellhole this place was, compared to the pristine forests from where Sundar was abducted. Elephants never forget! Surely, memories of his home must have haunted him.

We resumed our investigation and tried to find out where Sundar might have been tethered during the day. I had walked on the temple floor for barely fifteen minutes, but my feet began to burn, and I had to pour some water to cool them off. As we entered the main square, I suddenly tripped over an iron ring embedded in the stone. This was the place where Sundar had been chained and forced to stand under the scorching sun with little food or water, day after day for over six years. He would arrive at daybreak and stand in this very same spot until the sunset, amusing people, while raising money for the temple.

It was a poignant moment touching the ring where Sundar was shackled! I closed my eyes, feeling Sundar, and I remember praying, 'God, I want just one glimpse of Sundar. You are a merciful, God, so please help me to meet him.' I then visualized, whispering into Sundar's ears, 'I won't give up on you and hope you won't give up on yourself or your brothers and sisters who are being tortured relentlessly. Where are you? Please reveal yourself to me, just once!'

The drums and pipes of a marriage procession jolted me

back to reality. People dressed in colourful outfits danced and celebrated the union of two souls as I prayed for my union with my beloved Sundar. The next minute, a funeral procession with people wearing white clothes and hats walked past, carrying a dead body. One family was rejoicing, while another was mourning the loss of a loved one, within split seconds in the same place. What a stark contrast!

After wavering for several months about including Sundar's saga in my documentary, I was now determined to share his story. I turned to my driver and with firm resolve told him, 'We have to find Sundar.' And that we would not be leaving Kolhapur before seeing Sundar. Fortunately, since it was his family temple, he did not hesitate to return to the priest and elicit some help in finding the exact location of Sundar.

As expected, the priest directed my driver to a place that was familiar to him. Initially, I was a bit sceptical and wondered if this could be a trap. After all, it was because of people like me reporting to PETA that Sundar had been removed from the temple in the first place. By now, dusk was settling, and the skies were turning amber, pink, and yellow.

After an erratic drive on the main road, we were in for another treacherous drive in an industrial area. The oncoming traffic on a tapered and muddy pathway was relentless. Two scrawny and exhausted bulls dragged a bullock cart carrying a massive load of goods, and behind them were several motorcycles trying to get through the narrow lane. In the distance, I noticed two tall chimneys releasing dark, dense smoke, polluting the air. Suddenly, we found ourselves at the end of the rugged and dusty road. There wasn't a soul in sight! My heart sank. I thought that we had been duped. After having come this far, my hopes of finding Sundar began to dwindle. My convictions were being tested.

As the driver and I stood there hopelessly, a man riding a bicycle, with his mother on the backseat, appeared out of nowhere. My driver stopped him and inquired about Sundar. The man told us that we had missed a turn and were only five minutes away from the place where Sundar was. So, we drove back, turned into a lane as directed by the cyclist, which led us into a school compound. Four watchmen were alerted. However, my driver disarmed them by speaking their language and informing them that he was from the same village. After he explained our mission, they directed us to the rear of the building.

A few meters away, we noticed a tin shed. Beneath it, a tall grey figure swayed from side to side. As we approached the tent, a cement structure had several posters of Sundar glued to it. Opposite were a couple of large hay sheds that looked like chicken coups. And there he was! The stunning Sundar! After almost fifteen hours of driving into the boonies, we had finally come face to face with this media sensation.

It felt surreal to be in the presence of this special elephant. I sat in a car for a few moments, trying to come to grips with my swelling emotions. I had to be strong enough to gather important footage, so I could present the story of this brave warrior. I eventually stepped out of the vehicle and approached Sundar. After all the twists and turns, not knowing if we were on the right track and having put ourselves in harm's way, I was now standing before the magnificent animal. However, his shed was barricaded.

Sundar had a roof over his head, apparently to protect him from the sun. But it was made of tin sheet, which would only intensify the blazing heat. Thankfully, there were no walls around the tent. At least, the breeze would cool him off a bit. A vast barren land stretched as far as the eye could see. It was owned by the politician, the culprit who had bought Sundar and donated him to Jyotiba Temple.

As I was taking in the surroundings, a watchman and a woman with her two teenagers came running aggressively towards our car. My escort had warned me of the inherent dangers in the situation. But unfettered by the drama, I walked past the barricade. Now, I was just a few meters away from Sundar. I stood there silently, looking at him. I then turned towards the mad people who were running at me and smiled with my palms joined in a gesture of namaste. They calmed down, smiled back, and said namaste. I then struck up a conversation, explaining how much I loved elephants.

I turned to Sundar and called his name. In a gesture of acknowledgement, he lifted his trunk and playfully rubbed his forehead. It was mind-blowing. It seemed as though Sundar knew I was there for him, that a world out there was looking out for him, that he was loved, and that help was on its way. I suddenly noticed the sun setting behind him, its remaining rays shining through a tuft of hair on his head. He was still only a young bull.

In the wild, the boy elephants hang around with their mums until they are about fifteen years of age. Although they may try to show off their machismo among their friends, they still require the presence of the matriarch for guidance. But poor Sundar had been separated from his family when he was just eight years old.

The guard said, 'He likes you,' to which I replied, 'I feel him.' That was when I pulled out my camera, which was hidden behind my back until this time. Instantly, the woman began to yell and scream. The guard blocked my camera and pushed me, asking us to leave the premises. He said they had a court order that precluded the taking of photographs by any visitors. I respectfully left the area and stood near the barricade. Instead of filming Sundar, I filmed the chicken coop, the people, and of course, the shed. The three times that I turned towards Sundar to film him, the guard blocked my camera. I still managed to capture a few shots, but unfortunately could not get any close-ups. The owner's wife called a few men to intimidate us. My escort said we should leave, but it was hard to do that. My heart was heavy!

As we drove away, I turned back one last time to look at Sundar, until he faded away into the distance. As we veered off the muddy road and were just about to enter the highway, my driver noticed in his rear-view mirror a few motorcycles following us. They were deployed by the owner's wife. I asked my driver to pull over, but he was apprehensive. I sternly repeated the request, at which point, he obliged. The bikers surrounded our car. I looked at them and smiled and said, 'Namaste.' They smiled, revved their bikes, and rode away.

We made it back to the highway and to Mumbai, arriving at 4:00 a.m. the next day in what had been a non-stop journey of twenty-two hours. But there was no rest for this weary soul. After tossing and turning in my bed with the images of Sundar still fresh in my mind, I was up within a couple of hours. That day, my mum had invited me for lunch. Later that evening, I flew back to Toronto.

A couple of days later, I learnt that the High Court of Bombay had finally ordered Sundar's release and also mandated that he be transported immediately to the nearest sanctuary. But the defiant politician, who had donated Sundar to Jyotiba Temple, refused to obey the commands. Instead, he filed a Supreme Court order against the High Court. The politician's appeal was quashed by the Supreme Court, which stayed the high court's orders. This should have been a victory for Sundar. But still he was not released, the excuse being that he was unfit to travel.

A government veterinarian had examined Sundar and discovered a large wound on his right ankle, proving that his torture and abuse had not stopped. With each passing day, his suffering only intensified. Finally, after months of emotional roller-coaster rides, justice and common sense prevailed, and Sunder was freed. This landmark victory set a precedent for all temple elephants. Collective efforts had triumphed, even over a very powerful politician.

Sundar's happy ending would prove to be the only positive story in my entire film. As I reflected on the course of events pertaining to this resilient elephant, I felt an invisible hand in all of it. Hadn't it been a most profound coincidence that my brother's driver happened to be from the same small town, and visit the same temple where Sundar was held a prisoner? My prayers to meet Sundar had been granted, with the generous and unbridled help of complete strangers along the way. I am still in awe, trying to comprehend the unfolding of so many synchronicities. This experience, like the others, only strengthened my conviction that I was being used as an instrument of the divine will. And I wondered where the next instalment of my mystical odyssey would take me. In the ensuing weeks and months, I would be rinsed like a cloth every step of the way as some of the people I met triggered my inner demons.

Chapter 11

Putting the Film Together

Now that I had all the footage and interviews, the next piece of the puzzle was to find a suitable post-production team. This would be a difficult task. After returning to Toronto, my first task was to sift through almost two hundred hours of footage and describe each and every shot by the second. I also needed to transcribe the interviews word for word, select powerful and relevant shots, and map out a story. With no assistant producer to help, I turned to my media colleague, Bill Hutchison, who had introduced me to crowdsourcing. He was also a media professor at one of the community colleges, and he found me an intern who created a shot list.

Meanwhile, it dawned on me that I had never done anything like this before. As a nature and wildlife reporter, I had produced three-minute stories for the daily newscast. After transitioning into film-making, I had produced short half-hour documentaries, but never feature-length (over an hour-long) films. Even so, nothing could have prepared me for a challenge of this magnitude. I quickly discovered that I needed help. Thus, I reached out to an editor who had freelanced for me for nearly six years.

Maneesh was only twenty-two years old when I had first met him in 2009 through a mutual friend. Given that he was of Indian origin like myself, he and I connected right away on a cultural and personal level. I hired him immediately to help me with the short documentary films I was making about Bermuda's nature and wildlife at that time. He told me that his mother had died when he was a young boy and, as a result, he was raised by his father. Naturally, I felt a deep maternal bond and empathy for him. During our two to three years of working together on my Bermuda films, I would often feed him homemade meals and inquire after his health and overall welfare. I still remember editing with him, often non-stop, for eighteen to twenty hours a day, sometimes until 4:00 a.m., in order to meet deadlines. After working such long hours, we would often become quite punchy and laugh non-stop at the silliest of jokes.

Maneesh did not have an office space to edit nor a professional audio-recording booth. But we somehow managed to produce a six-part series, *Bermuda—Nature's Jewel*, which received the best environmental awareness award from the Bermuda Audubon Society and was commended in the upper and lower houses of Bermuda's parliament. A shorter version of the series would be eventually featured on Discovery Channel Canada.

So naturally, after I returned from Kerala with such a powerful story, I was inclined to invite Maneesh to edit my film on elephants, *Gods in Shackles.* His initial job was to organize all the footage that Tony and I had shot in India, which involved finding the best shots and labelling everything. Although Maneesh had a full-time job, he was quite keen to come on board. He agreed to work evenings and weekends to be a part of the team. By now, he had secured a spacious edit-suite in which we could work. He'd even bought a brand-new video monitor. We planned to begin editing in the autumn of 2014, leaving enough time for scripting the story.

Most good documentaries have a written outline or overview that is followed as the film is constructed. Having been a journalist for almost two decades, I was leaning towards writing the script myself. But I felt a bit unsettled by the prospect. So, I turned to a friend and mentor, Paul Lewis, a veteran media executive. He suggested that it would be best to hire a scriptwriter as he felt that I was too close to the story. With his extensive media connections, he graciously offered to ask around for such a person and help me fill other key positions on the film-making team.

Paul and I had met in 2008 through a media colleague as I was wrapping up my contract at the television network in Bermuda. At that time, he held the top job at the Discovery Channel Network in Canada. I remember having spoken to him for forty minutes the very first time I had phoned him from Bermuda. Paul and I shared many things in common and chatted effortlessly about several topics. These included the twenty-four-hour news cycle, the necessity of objectivity in news reporting, and, in particular, our love of nature and wildlife. We felt a sense of familiarity with each other from the get-go.

It was in late January of 2009 when I had returned to Toronto for a brief visit that Paul and I met at a restaurant for the first time. It felt like a blind date, awaiting his arrival, as I sat at the table, sipping my fruit juice. In a few minutes, an unassuming medium-sized man with short brown hair and a boyish face approached my table and introduced himself. After a mild handshake, Paul dove right into his usual style of chatting and sarcastic sense of humour. He explained how the cut-throat industry and competition for high ratings had kept him busy. He also said that the network is always looking for new content. So, we decided to stay in touch and explore the possibility of featuring some of my short films on a popular show, *The Daily Planet*.

I returned to Bermuda in mid-2009 to direct and manage the Bermuda Environmental Alliance (BEA). During the five years of managing the BEA, I shifted back and forth between Toronto and Bermuda every other month as the cost of living in Bermuda was formidable. Besides, I had to return to Toronto to edit with Maneesh the six-part series.

The show would turn out to be extremely comprehensive and go on to win several awards. I certainly couldn't have pulled it off without Paul's guidance.

Over the years, we had become the best of friends who could always count on each other for moral and emotional support. In the six years that I'd known him, he offered his profound wisdom as well as feedback on my scripts. He also taught me how to communicate with difficult people. He was articulate, and I admired his vocabulary and literary skills. One of the most amazing things about our friendship was that it was unconditional; we placed no expectations on one another, and there was no pressure. Even if we hadn't phoned each other or exchanged e-mails for weeks, or sometimes even months, we always picked up where we had left when we met again. Over time, our bond strengthened organically.

Fast forward to 2014. Given my background with Paul and how much I trusted him, it felt natural to ask him to become my media advisor for *Gods in Shackles*. After he had viewed the dramatic footage that Tony and I had shot in Kerala, he was blown away by it, and he agreed to advise me. He not only mentored me every step of the way, but also coached me on how to best recruit my entire post-production team. Paul was the candle that brightened my path throughout the production phase of *Gods in Shackles*. He introduced me to a writer named Digby Cook whom he had worked with at the Canadian Broadcasting Corporation, the CBC television network. I knew one thing for sure—anyone who has worked at that national news network would be an excellent writer. So, I was elated to meet Digby in early February 2015.

I vividly remember the first time we had met, he carried a black briefcase, as though he was attending an interview. Digby was six feet tall and had a crown of silver hair. His deep voice, sarcastic sense of humour, roaring laughter, and, of course, his love of Indian food pretty much set the tone for our ensuing conversation. He spoke his mind clearly and, as journalists, we shared similar perspectives and world views. For instance, presenting viewpoints of various stakeholders for objectivity, providing factual information, and avoiding sensationalism, among other things. We felt comfortable with each other, especially because of Digby's British upbringing, intricately connected with Indian culture. Trust and confidentiality were a given as we were both grounded in journalism principles. So, we cut a deal: I would cook Indian food and cover Digby's hotel expenses in India if he provided his services for half of his regular fees. It turned out to be a sweet arrangement for both of us.

After some hot chai and Indian snacks, we screened the footage on my laptop. Digby was clearly gripped by the dramatic videos and he immediately knew that the powerful images would dictate the story. Our only job was to provide the proper context for them. I had written articles for the Huffington Post on the atrocities against elephants, which we used as a blueprint for certain sections of the script. Digby also understood the cultural context that was part and parcel of the elephants' issues. His challenge was to create a lively script that would resonate with people everywhere.

One of the first things he suggested was that the film should be a personal journey of mine. But I wasn't too comfortable with the idea as I feared that my presence could deflect attention from the elephants' suffering. So, we turned to Paul for advice, and he concurred with Digby's assessment. I was still uncomfortable drawing attention to myself, but I went with the flow and agreed to follow their recommendations.

Given his decades of writing experience, I was a bit reluctant to instruct or direct him. Deep down, I was feeling intimidated. But I realized that I was being stretched out of my comfort zone. Once again, I turned to Paul for advice. He pushed me to articulate the situation, replete with its cultural sensitivities more clearly to Digby.

The following week Digby and I met to review more shots, given that I had identified some missing beats in the script. This time I openly shared my vision for the script, and we agreed to allow the videos to tell the story, rather than adding a tone of sarcasm as an overlay. As I transcribed the last fifteen interviews from India, Digby worked on at least thirty-seven revisions to the script. Finally, we found our flow and the tone we wanted for the documentary. Overall, we made a great team and, most importantly, our work together had become enjoyable. Actually, we looked forward to our work sessions because we had fun and complemented each other's writing skills.

Now, it was time to edit the film. While Digby and I prepared the script, I had expected Maneesh to have organized the two-hundred-plus hours of footage. So, I called him and left a voice message to plan our editing schedule. For weeks I did not hear back from him and I began to feel uneasy. But Maneesh eventually contacted me and together we created a tentative schedule with the understanding that most of the editing would take place on the weekends. However, when it came time to begin, Maneesh postponed the session. Apparently, the workload of his day job had increased. Autumn of 2014 had passed, and then winter set in, but still, Maneesh had nothing to show me.

Having worked with him for more than six years, I had convinced myself that he was the perfect person for the job. But in February 2015, as the script was being generated, a bombshell was dropped on me. Maneesh was facing a health crisis, and he broke the news that he could not edit my film. Be that as it may, I had a hard time letting him go. In fact, for several weeks, I was still holding onto false hopes that, perhaps, things might improve, and he would stay on to edit my film. But eventually, I had to swallow the bitter truth.

The deadline for the movie release was early December 2015, and with less than ten months left, I had to return to the drawing board to find another editor. Fortunately, Maneesh did not leave me stranded. He created a job posting on the Internet and applications poured in from highly experienced veteran editors.

Amid all the resumes for the editor's position, one from a music composer dropped into my inbox. Janal Bechthold had read the ad and sent a compelling e-mail expressing keen interest in composing music for my film. I was desperately looking for an editor, and a music composer was not even on my radar at that particular time. But something prompted me to respond to her. We connected via Skype and I openly told her that this was a low-budget production, but she was fine with that. I can't explain why, but during our conversation, it slipped out of my mouth that as a young girl, I loved singing classical Indian songs. However, after moving to Canada, I'd abandoned everything pertaining to Indian culture.

Then I did something bizarre—I offered to sing to her via Skype. Not surprisingly, Janal was taken aback by my preposterous proposition, but in a hesitant voice, she encouraged me to go ahead. Sure enough, she ended up liking my voice and the songs. She then asked me to come down to her studio in the Distillery District of old Toronto where I sang a few more songs for her. In the days ahead, we booked a couple of recording dates for four of my classical Indian renditions, and the rest, as they say, is history. After we got to know each other, Janal confessed that when I offered to sing on Skype, she rolled her eyes and thought I was strange. We had a hearty laugh at that!

It was at this point that Janal contacted her mentor, a world-renowned and international award-winning music composer, Donald Quan. The two of them in conjunction with two other Indian musicians composed an entire library of music exclusively for *Gods in Shackles*. My performance turned out to be a blessing in disguise because after we had launched the movie trailer, I was targeted by certain groups who tried to frame me as a 'waspy Indian woman' disconnected from her cultural roots. They said this to undermine my observation that the holy scriptures were being perverted in light of the abuse that the temple elephants suffered.

Meanwhile, I had decided on three very suitable editors. After consulting with Paul, I selected Rob for an interview. It was in early February 2015 that we met at the popular Tim Hortons. It wasn't easy to miss Rob as he entered the coffee shop with his large glasses and crown of long, curly hair. As we chatted, I sensed his high energy level and enthusiasm. He was one of those hyper-serious-techno-geek types but seemed flexible. And this was an important rationale for hiring him. After signing an agreement, Rob went off on a pre-planned vacation for ten days.

I next had to figure out where we would edit the film. Renting an editing suite was out of the question as it would cost a minimum of twenty-five hundred dollars a week, which wasn't budgeted for. Once Rob was back from his holiday, he would drop by my home office every day with his large Apple computer installed with robust editing software. Within the first week I realized that Rob was an unbound creative type who would not adhere to a set schedule. On the other hand, I had always been rigid about a regular routine. But he was a fast editor and having edited popular reality shows, he was used to grabbing random shots and slapping them into a narrative. However, a project as sensitive as *Gods in Shackles* required meticulous and reflective editing. The movie was culturally explosive. Thus, I had to constantly remind him about using videos thoughtfully in a manner that would move and empower the masses. All this delayed the editing of the film, which, in turn, slowed down the overall post-production process.

The entire winter went by rather quickly. But even after almost two months of working with Rob, I was still unable to get over Maneesh. I was feeling deeply concerned about his health and well-being. I missed sharing food with Maneesh, our silly jokes, and just being with him. My only solace during these difficult times was Mother Nature.

Soon, the spring came into full bloom. The days became longer and brighter. And walking in the woods was the only outdoor activity that enlivened me; the rest of the day, I was inside the confines of four imposing walls. There was something lively about a wilderness pond that I frequented. Watching the little insects, dragonflies, frogs, and even a tiny turtle basking in the sun comforted me.

One day, I spotted a goose sitting on a small mound across the pond, content and unperturbed by my presence. Days went by and the goose remained in the same spot. Then it dawned on me that she had laid eggs in the mound and was protecting them with the warmth of her body. She somehow knew that this is what she was supposed to do at this time.

I visited the goose daily for the next four weeks. Rain or shine, snow or wind, she would not abandon her precious eggs. Despite isolating herself onto a tiny island of sorts near the pond, the goose was constantly threatened by people walking their dogs on the ground, birds of prey hovering above her, and other predators at night. And yet she remained undeterred. She would occasionally stand and stretch and preen herself, plucking out her feathers to keep the eggs warm. But other than that, she sacrificed all her comfort to guard her precious offspring. This goose was intelligent, instinctual, and intuitive. She embodied Mother Nature.

Similarly, elephants fiercely protect their babies. One of the scientists had even documented a wild elephant protecting a dying baby who didn't even belong to her. They know intuitively what to do and how to handle challenges as they are acutely aware and deeply connected with their origins. By reclaiming my own intrinsic bond with the creatures of the earth, I was able to discover tremendous stores of resilience within myself.

The seeds were sprouting. I simply needed to allow them to sprout, just as the goose was allowing her goslings to hatch in their own time. The goose was birthing her goslings, and I was birthing my film! Everything else would flow—nature would handle the rest. Was the goose trying to convey that message to me?

Perhaps the message was for me to remain unperturbed by what was going on in my life—or not going on—and let go of situations and relationships that were not meant to be. There must be an intelligence that provided the goose with the knowing, strength, and resources to serve her purpose of giving birth to her goslings. If the universe supplied her with resources, strength, and knowledge, why was I fretting? No doubt the universe would supply the same resources to me.

Through unusual circumstances, the seed to produce *Gods in Shackles* had been planted in my mind. Serendipities had unfolded and resources had been provided. I now needed to muster up my strength and release those things that no longer served my purpose. I honoured my special bond with this precious goose and learnt many lessons from her—surrender, patience, dedication, commitment, and being open and receptive to the resources and people that were surrounding me. My only job was to focus on serving my purpose.

Finally, after nearly thirty days, on the morning of 27 May 2015, four adorable goslings, together with mother goose and the gander, were swimming in the pond. I will treasure the two hours I spent with them for the rest of my life. One adventurous gosling, seemingly the boldest, ventured into the pond as others sought shelter beneath mother goose's wings. The gosling then scurried back up the mound, balancing itself with its tiny tender wings and then pushed aside the other goslings to stay closer to its mother's bosom. Somehow, they seemed to remember the warmth and security they had felt in the egg, stored in the mother's womb. How did these babies know to scurry beneath their mother's bosom? How did they know that their mother protected them by keeping them in her bosom before they'd hatched?

Soon after this, the weather became drastic. Toronto was hit by two continuous days of thunderstorms with a tornado warning in effect. I worried for the family of geese. Two days of torrential rains; where would they hide? As soothing as the natural world is, it can turn dangerous on a dime.

After the storm subsided, I returned to the pond. But as I had feared, it was empty! What had happened to the birds? The goslings could not have swum all that far away. I thought of the worst. Perhaps they had been swept away by the flood's waters. I desperately searched for the family everywhere, walking as far as I could on the banks of the Don Creek. I was clinging onto the hope that I might find them somehow. Sadly, I did not. I returned to the pond, and as I looked around, the place looked lifeless. The mound where mother goose had sat for more than four weeks now only displayed remnants of her feathers in the nest. This search operation went on for the next four days. But every time, I returned home with a heavy heart. Still, I somehow felt their precious presence near me.

On the fifth day, I decided that I would make one more attempt at finding them, but it would be for the last time. I was tired of being disappointed. Again, this time, no luck! I gave up and turned around to head home. Immersed in my thoughts, I walked the trails with a sense of emptiness, worrying about these birds.

Then, suddenly, I noticed some movement on the other side of the creek near a bridge I was approaching. I picked up my pace with a bounce in my step and hope in my heart. Lo and behold! There they were! A happy family of six—mama and papa goose next to their four adorable goslings. They were just a yard away from me, happily feeding on some algae. They had survived one of the worst storms of the decade.

Around the same time, my film's rough cut had been completed. The goslings had been born and so had my film. This symbolized a few things for me. My creation would survive and thrive through the challenges, just as the goslings had withstood the thunderstorm. As the geese family had been given the resilience and wisdom to cope with thunderstorms, I, too, would receive the strength and insight to withstand the storms of my life.

I worried for their safety, yet they had cleverly sought shelter from the storm. After the weather had cleared, they were back, happily gallivanting, busy being the goslings they were meant to be.

In order to survive the storms of my mind and clear the cloudiness, I needed to take breaks. Second, when I accepted the geese's disappearance and surrendered my concerns with prayers for their protection, they showed up. What a great omen this was!

After my reunion with my goose family, I was ready to embark upon a monumental journey. In two days, I was due to return to India on a fact-checking mission to ensure that the facts presented in my documentary were accurate. I had worked on the rough cut for three straight days with Rob—until 4:30 in the morning each time—and we had finalized it. The next day, Rob burnt a Blu-ray and delivered it to me. This was the disc that I would take to India with me.

But as last-minute challenges can and do often happen, the disc turned out to be corrupt. Rob re-edited and re-synced the rough cut's sound and video just twenty-four hours before I was due to depart. I flew out that same evening to Bangalore with the *Gods in Shackles* rough cut, terrified that it would be scrutinized meticulously and critiqued by the experts. I was even preparing myself mentally to re-edit the entire film.

Chapter 12

ELEPHANT ALLIES

Throughout my flight to Bangalore, I kept worrying about what others would think and how the experts would evaluate the film's rough cut. I worried about the technical glitches and everything else that could go wrong. Fear gripped me and I began to panic as the plane took off. I ordered a glass of sparkling wine, trying to relax. I arrived in Bangalore at midnight and took a cab to my hotel in the wee hours, where I managed to sleep for a couple of hours.

In addition to screening the film's rough cut for feedback, I had returned to India to interview two very special people for it, about whom I share in the next chapter. Suparna, who had been interviewed for my film in Bangalore, had organized what would be a humble screening at her downtown apartment in Bangalore later in the week. I called her early the next day to check up on the arrangements. We had lunch together, and I shared my fears with her. But the gentlewoman showered me with words of comfort, reassuring me that everything would be fine.

All too quickly, the day I dreaded arrived. Some of the key people featured in the film would be in attendance at the screening, including Dr Raman Sukumar, whom you might recall I had met on my first trip to Kerala. After everyone was assembled and before the screening began, I made an appeal, asking everyone to make notes and share their thoughts after the movie ended. For almost two hours, there was pin-drop silence as everyone's eyes were glued to the TV screen. Although they had been aware of the unimaginable atrocities that were being perpetrated all over India against the elephants, they were unprepared to witness the documentation of these abuses in such a revealing manner.

After the screening, Suparna announced that she had launched a Supreme Court case against Indian states that exploited elephants. One of the audience members whom Suparna had invited to the screening was the then Animal Welfare Board of India (AWBI) secretary, Vinod Kumar. He said that he was deeply moved by the film and that he knew all too well what's happening in Kerala because he hailed from that state. I asked if he would give me an on-camera interview and he agreed without any hesitation. So there and then, I set up my camera gear—my perennial friend—and the secretary spoke, touching on key legal points. He did not mince his words but crafted them very carefully.

Others provided constructive suggestions as I mixed and mingled with them later on. They said they were impressed by the accuracy and objectivity of the film and the rich sound effects and narrative. Their feedback meant the world to me as it eclipsed all the technical glitches that I feared would be magnified onscreen. It was after all just a rough cut. I could now rest peacefully after such an overwhelmingly positive feedback. I was incredibly uplifted by the reaction to the film, and, that night, I slept like a child after having spent several sleepless nights before the monumental screening.

Following the fact-checking that I did in Bangalore, I travelled to Kerala. The first place I visited was Lakshmi's home. Precious memories flashed back vividly as we drove on the same road where Tony and I had filmed a particularly poignant scene. It had been the last day of filming in May 2014, and we were making what would be our final journey to the temple. It was peak rush hour and as Lakshmi was walking through the congested traffic, she spotted a dead cat on the road and so did I. She paused for a few moments in reverence and stepped aside carefully. She then circled around its body, as though mourning the loss of a precious life. Meanwhile, the oblivious drivers drove over the cat's little corpse.

It has been proven that elephants grieve the loss of their loved ones in many ways, including touching the remnants of their dead. In this instance, it was a cat. But regardless, Lakshmi paid homage to another species in her own gentle way. Despite her own suffering, the reverence she displayed for the dead cat will remain inscribed in my heart forever.

Now, I was on my way to reunite with this beauty. On my drive to see her, I phoned one of her handlers who told me that Lakshmi was sick. When I probed deeper, he sounded a bit apprehensive, suggesting that I should find out on my own. I immediately knew that something was very wrong and asked my driver to take me to her place straightaway. As soon as our car drove through the two massive gates into the large compound, I ran toward her tethering spot. There she was, swaying restlessly, her left eye swollen shut, tears streaming down her face. Lakshmi had the most gorgeous pair of eyes. Her eyelashes were so long and curly that it looked like she had worn mascara. When she slowly opened her eye, I noticed a white scar that had replaced one of her hypnotic honey-brown eyes. To witness her struggling to open this eye utterly ravaged my soul.

One of the other handlers told me that her main handler was responsible for the injury. Apparently, Lakshmi had eaten a bag of vegetables that he had left unattended at the temple. In a fit of rage, he beat her mercilessly with his vicious weapon and one of his blows hit her eye. It was painful to watch Lakshmi using the tip of her trunk to rub her traumatized eye, trying to soothe herself.

Fortunately, I had carried my camera, and as I began to film her, it suddenly dawned on me that, perhaps, the film's climax was unfolding right in front of me. I spent the entire day with Lakshmi. Later in the afternoon, her main handler approached her and commanded her to lie down. But Lakshmi was terrified, and she vehemently refused to obey him, trumpeting and urinating in distress. The handler walked away and returned with his vicious weapon. Despite the tightness of her shackles, Lakshmi somehow managed to kneel down and then rolled her body in her urine.

At this point, four mahouts came dashing onto the scene. Two of them held her down, as the third one poked her eye with a filthy hand that was caked with dirt. I yelled at the mahout and told him to wash his hands. He ignored me. I then informed the supervisor and asked him to order the mahouts to wash their hands. He did, and the mahout reluctantly complied, barely dipping his hand in a bucket nearby. By now, the mahouts were getting agitated at Lakshmi as she refused to open her eye. Her handler began to yell at her. She panicked and closed her eye tighter. Finally, the third mahout sat on her head, forced open her eyelids, and popped out her eyeball. Her handler then poured some eye drops into the eye socket.

I don't know how I held the camera steady through the entire ordeal. I don't know how I became aware of my emotions and managed to control them in order to document the atrocities against this docile animal. I could only imagine how harrowing this must have been to my beloved Lakshmi, as she laid there helplessly with her eye closed.

As soon as I'd finished filming, I phoned a renowned veterinarian, Dr Jacob Cheeran, whom I had met at the veterinary university. I narrated to him in a shattered tone exactly what I had seen. He told me, as a matter of fact, that the treatment would only exacerbate the injury. He explained, 'The basic medical principle is that if the therapeutic hazard supersedes the disease/injury then such treatment should be avoided or ceased immediately.' In other words, if the treatment is more dangerous than the injury then it would be best to leave the afflicted area to heal naturally. I immediately phoned the mahouts' supervisor and told him what the vet had told me. I also explained that the vet should be treating the elephant, not the handlers.

However, during the five days that I visited Lakshmi, Dr Giridas, the government vet who was featured in *Gods in Shackles*, never showed up. It revealed how little he cared for the elephants under his care.

My mission in Thrissur was nearing its end, and I absolutely had to visit Lakshmi to say my goodbyes. But despite her eye injury, she had been taken to the temple to perform the ritual circuit. So, in her absence, I began to feed fruits to the other two elephants in the compound, Ayyappan and Shiva Sundar, both of whom are featured in *Gods in Shackles*. Although Shiva Sundar was in his peak mating season, frustrated, and throwing rocks at his handlers, he allowed me to feed him balls of rice and lentils. Ayyappan also received me warmly, trumpeting, as though telling me, 'I know you have fruits. Now come and feed me!'

As I began walking towards Ayyappan I heard the haunting sounds of shackles echoing from a distance. I knew it was Lakshmi. I was standing to the left of the compound, but she couldn't see me due to the loss of vision in her left eye. Still, she sensed my presence and stopped. She was trying to reach out to me with her trunk. But the mahout gave me a dirty look and dragged her away. I followed them with the bag of fruits I had brought for her. Ignoring the handler, I fed Lakshmi pineapple as well as lots of guavas, watermelons, carrots, and her absolute favourite—bananas.

After a few minutes, Lakshmi's abuser and I came face to face. He tried to avoid me, pretending that I didn't exist. But eventually, I approached him and politely asked how he was doing. Although I realized that a person is considered innocent until proven guilty, I could not let him off the hook. I confronted him and asked on-camera why he had hit her. He vehemently denied it. He then went to the extreme of offering to dip his hands in boiling oil in an attempt to convince me that they wouldn't burn as he was speaking the truth. Obviously, I rejected that offer. He eventually confessed to having hit her on her legs because she 'stole' his vegetables. However, none of his rationales mattered for I had all the materials I needed on film to expose the truth. Frankly, I am unsure where I conjured up the strength to stay calm and collected with him as emotions were running high.

It was now time to say goodbye to these precious animals. I talked to Lakshmi and told her that I would be back soon. Pearls of tears flowed down her right eye. Although elephants don't have tear glands and can't control the flow of tears, I knew in my heart that she was crying. I sensed that she felt my pain and distress that I was leaving her in the clutches of these cruel men. It was indeed a painful departure, heartbreaking to look at this majestic animal whose gorgeous honey-brown eye had been now replaced by a white scar.

After returning to my hotel room in Kerala, I reviewed the footage. I selected still shots of Lakshmi's abuse, including the way medication had been forced into her eye. I then filed a report with the Animal Welfare Board of India. An investigation was launched against the four handlers who had ignored proper medical protocol and the government vets were promptly dispatched. They discovered that Lakshmi was partially blind, with a prognosis that she could become permanently blind. In the ensuing weeks, I was informed that the temple authorities had fired the abuser and that she would have a new handler. Although the new one seemed compassionate, this didn't speak to the issue at hand—should elephants ever be held captive?

I had gathered ample evidence to expose the horrific abuse of elephants, but there were still key beats missing. I needed the voice of a credible animal rights advocate and a religious authority to rally behind the elephants' cause. Fortunately, they would prove to be close at hand.

Chapter 13

The Crucial Missing Links

On a cool evening in the spring of 2015, when I was back in Toronto, I googled Sugathakumari and landed on a web page loaded with information about her. As Suparna had told me during my interview with her, Sugathakumari was Kerala's beloved poet laureate. Among other things, this admired activist had received the presidential award just a few years back. Amazingly, I also found her phone number. Now I was faced with the more difficult task of contacting this celebrity. It was a bit nerve-wracking. Yet, somehow, I mustered up the courage to pick up the phone and dial her number. The phone rang and lo and behold, a kind and gentle voice said, 'Hello.' My heart melted, but maintaining my composure, I asked if I was speaking with Sugathakumari. She said, 'Speaking.' I could not believe that she was so easily accessible! I continued to breathe deeply during an engaged conversation that lasted for nearly twenty minutes.

The intensity and passion in her voice reverberated through the telephone, travelling oceans across, Kerala to Canada. I told her about my documentary and asked if she would be willing to be interviewed for it. Without any hesitation, she agreed. After I hung up the phone, I sat on my sofa and had to pinch myself to believe that this had actually happened. I had just spoken with one of the most revered women in India, and I knew that the meeting we had planned would be very, very special.

I was now in India to meet the legend, keeping the appointment we had arranged beforehand on that phone call from Toronto. I was both excited and nervous at the same time. Sugathakumari, whom everyone affectionately called 'teacher', lived in Kerala's capital city of Trivandrum. Through her evocative poems, she had been an eloquent spokeswoman all her life for Mother Earth and her creatures. As a prolific writer, she also shed light on the plight of abused women and ran a shelter for abandoned, abused, and mentally disabled women. She would be perfect to bring new perspectives and depth to my documentary, adding a very influential voice to the film.

Meanwhile, I had also heard about Akkeramon Kalidasan Bhattathirippad, a Hindu priest in Kerala who had spoken out against the exploitation of elephants on several occasions. He was a revered figure and the president of the state unit of a Hindu organization called Kerala Yogakshema Sabha (KYS). In many of his media interviews, this honourable man had publicly condemned the practice of exploiting elephants in religious and cultural festivals. He described it as an 'unnecessary practice paving the way to incidents of mishaps involving elephants and members of the priestly community as well as the devotees'. This priest would be an ideal voice for the elephants. But I had no way of contacting him.

In what would prove to be another powerful synchronicity, through unexpected twists and turns, an elephant activist, Rajeev, from Kerala contacted me via Facebook. He not only offered to introduce me to Akkeramon Kalidasan Bhattathirippad, but also to accompany me on the initial meeting with him. As it turned out, the priest lived between Thrissur—where I usually stayed when I visited Kerala—and the capital city of Trivandrum. So, it made sense to interview the poet and the priest on the same day. As mentioned in the previous chapter, in addition to screening my film in Bangalore, another main reason for travelling to India in June 2015 was to interview these two very influential people.

Early next morning, it was pitch dark when I stepped outside my Thrissur hotel with my camera gear and pages of questions that I had prepared for the interviews. Salu, my driver had promptly arrived, ready to embark on the long trip to Thiruvalla, a 165-kilometre road trip. As we drove through the empty roads, I began to film my sojourns. The air was already warm and humid. After gathering a few driving clips, I closed my eyes, trying to come to grips with the reality that was unfolding. I was on my way to meet two noble souls.

Just ninety minutes into our drive, the sun began to rise. It highlighted a silhouette of swaying palm trees framing a backdrop of magnificent amber and yellow skies. A flock of rambunctious ravens suddenly emerged from the trees' branches and took flight, ready to face the challenges of the day. The red ball of fire was dispelling the darkness with brilliant golden rays that revealed a sweeping curtain of clear blue overhead. My driver and I stopped for some breakfast at a restaurant that was a favourite of mine, where we devoured a delicious repast.

After driving for four more hours, we entered the small town where the priest lived. Our first job was to pick up Rajeev who had travelled here by bus from a nearby district. It was a delight to finally meet this tall, slender man who had sacrificed his Saturday by travelling two hours from his hometown to take us to this prearranged meeting with Akkeramon Kalidasan Bhattathirippad, the renowned Hindu priest. As we drove along dusty roads, Rajeev explained that this priest oversaw more than five hundred Hindu temples in Kerala. I had gathered from my own research that he was one of those very rare religious figures who dared challenge the status quo. His influential voice would be vital to our film. In honouring Hindu traditions, we stopped at a general store and bought some fruits, a coconut, and some betel leaves. These leaves are traditionally used during Vedic rituals, auspicious occasions, and, in particular, while offering up *dhakshina*—a small donation—to priests, gurus, and elders.

After a thirty-minute drive along bumpy and muddy pathways, we arrived at an ancient, stately mansion, replete with hand-carved wooden doors, stone walls, and a brick roof. It resembled my home in Palakkad where I'd grown up. On the verandah were a few chairs scattered around, and a swing made of a large wooden board suspended from the ceiling. After Rajeev and I had settled in, one of the housekeepers greeted us with some hot tea, as is customary in India, and indicated that the priest was finishing up a meeting. In the courtyard, native birds were gloriously singing away. A flock of egrets was devouring dung mites, despite the intense heat.

As I was sipping my tea, cherishing the natural beauty, a man of medium stature with a dark beard and dark hair rolled into a bun emerged from the mansion. He wore a white silk sarong and had a shawl wrapped around his torso. On his forehead, right between his eyebrows, was a red dot of *kumkum*—a symbolic representation of the third eye or intuitive eye in Hinduism.

As he approached us, it was hard to ignore the powerful aura emanating from him. I offered him the gift of fruits we had brought and prostrated before him, seeking his blessing. As mentioned before, growing up in a Hindu family, this practice of honouring the elders had become natural to me. I then explained the purpose of my visit and asked if he would be willing to be interviewed for my film. He instantly drove us to his Vishnu temple where a ritual procession was underway and granted me permission to film it. This was the same kind of ritual procession that Lakshmi participated in. But here, there were no elephants in sight. Thus, it was natural to assume that this temple did not use elephants.

But as I was filming, suddenly an elephant and two mahouts entered the camera frame. I was puzzled for a moment and paused momentarily, but then resumed filming. I captured shots of the procession and the elephant. After the temple ritual ended, I turned to the priest and got right to it. Why, despite his strong opposition to using elephants in temple ceremonies, did his own temple use one in this way?

In a deep, calm, and reflective voice, the priest explained that the elephant, whose name was Rajan, had been a fixture at the temple before he, as the priest, had begun to preside over it. Prior to this priest's arrival, Rajan typically participated in various rituals of the temple ceremony—the circuits—in much the same manner as Lakshmi and the other temple elephants did. In this, Rajan had been adorned with heavy medallions and forced to carry a large plaque, honouring the Vishnu temple's deity on his back. However, after Akkeramon Kalidasan Bhattathirippad was ordained as the new priest, he ended these harsh practices. Instead of the elephants being forced to carry the idol, people carried it or dragged a chariot with the deity in it during his temples' ceremonies.

He explained, 'People are now carrying the idol. There is no need to have the elephant carry it, no need for it at all. They now just walk along as part of the procession in the same way that a human being does.' In this, the Vedic traditions lived on, proving that contrary to the claims of many religious institutions, it is not necessary for elephants to be an integral part of the rituals.

As my interview with the priest progressed, he quoted a poignant Hindu scripture from the Advaita Vedanta: '*Brahma Satyam, Jagath Mithya*', meaning 'the only universal truth is the *Brahman* (divine), the material world is an illusion'. Referring to the sacred texts of Thantrashastra, the priest argued that in no way did they advocate the use of elephants to parade the idols, asserting that elephants had never been a part of the ancient Hindu ceremonies and rituals. Instead, he said, the idol should be carried by the priest or placed inside a man-made chariot that illustrated other forms of animal life such as the bull, peacock, eagle, or lion.

However, he explained, humans have corrupted the meaning of the holy scriptures, twisting and distorting them in the name of culture and religion, by leasing out elephants for large sums of money. He went on to boldly denounce the veterinarians who provided fake fitness certificates for even ailing bull elephants and those who were in their annual musth cycle, posing a serious public safety threat to people. He also confirmed that the temple elephants tended to frequently run amok as a result of deprivation as well as the unbearable stress caused by the massive crowds, the bombardment of fireworks, and the scorching heat that they were regularly forced to endure.

After the interview, I needed some cutaway shots and without realizing that priests aren't allowed to sit on the barren floor, I asked this holy man to sit on the ground and assume various meditation postures. But without uttering a word, the priest sat on the granite floor with his eyes closed for a few moments. Entering into a deep trance, he was enveloped with peace and calm as he allowed me to film him beneath the scorching sun. This humble man embodied the Hindu philosophies. My encounter with this beautiful spirit in and of itself was a divine unfolding, and it certainly taught me humility. His untainted reflections, which I had captured in my camera, could potentially change the way the temple elephants in India were perceived.

By this time, the sun had reached its zenith at the midheaven of the sky. A shot of cool air ruffled the leaves of the peepal tree, commonly known as sacred fig and scientifically termed *Ficus religiosa*, beneath which we were standing. I took a deep breath, feeling a sense of profound fulfilment. Before leaving the temple, I prostrated before Lord Vishnu's altar as a way of expressing my deep gratitude for His divine guidance and for leading me to such fruitful encounters. For the first time in a while, I felt somewhat hopeful for the temple elephants of India.

I packed up my camera gear and Rajeev, my driver, and I went to a nearby restaurant for a hearty vegan meal, which turned out to be quite the elaborate lunch. *Sadhya* is a decadent traditional lunch that features at least a dozen dishes and three desserts on a banana leaf. After devouring the authentic Kerala meal, it was time to part with Rajeev, at least, for the time being.

My driver and I were now on our way to Trivandrum to meet Sugathakumari. We arrived late in the afternoon at her women's centre, which she had named *Abhaya*, meaning 'fearless' in Kerala's native language. Indeed, this poet laureate boldly used words and language as weapons to expose the atrocities against vulnerable beings—human and non-human. As soon as we arrived, I was taken into her office where it felt only natural to hug this medium-statured woman in her eighties, and she received me warmly, patting my back. After a few moments of small talk, she held my face in her palms and looked at me with her piercing eyes for a few seconds. Her prominent silver-grey hair and astute gaze spoke volumes. Then she said, 'Sangita, for decades I have been speaking out against these cruel practices. How shameful that they are using the name of God to inflict suffering on these innocent animals. God will only curse them, not bless them for their crimes against nature.' Her eyes welled with tears, as did mine.

My initial plan was to interview her in her office. However, after surveying the surroundings we felt it would be best to interview her outdoors, in serene, natural settings. So, we followed her as she was driven to her residence in a quiet and peaceful neighbourhood, a perfect place to interview her. By now, the sun was dimming. Given that the lighting had to be just so, I hurriedly set up my camera and began to pepper Sugathakumari with rapid-fire questions. But the wise poet responded reflectively, taking her own sweet time. When I realized that there would be no rushing this woman, I took a step back.

Sitting right beside her, I looked intently into her eyes and proceeded to ask her many questions, some personal. Without any hesitation, she answered every single one of them with full transparency. I decided right then that after making the film—which would be in English—I would have to produce a Malayalam version of it as well. So, I asked this amazing woman if she would consider translating the film from English to Malayalam and narrating it. Again, she promptly agreed. After enjoying our tea and some more small talk, my driver and I said our goodbyes and quietly took our leave.

The relationships that began on that fateful day with these two noble souls continue to strengthen with each passing year, blossoming into a trusting and respectful friendship. The additional perspectives that they had articulated about the temple elephants of India would not only afford the film their revered voices, but also form the part of its tragic climax.

Chapter 14

The Dark Nights of My Soul

Sharing this tragic story would do Ayyappan's memory justice. A handsome bull elephant, he was ten feet tall with dark grey skin, stunning tusks, and a sparkle in his light brown eyes. When I first met him in May 2014, he instantly captivated my heart and soul. As I stood near him, Ayyappan displayed his true nature by playfully curling up his trunk and rubbing his forehead with his mouth wide open for the goodies that he had already sniffed out. It didn't take him long to stretch his trunk and take the bananas I had proffered. These he quickly tossed into his mouth and devoured almost instantly.

I stood there gazing at Ayyappan, completely hypnotized by his presence. As he began to shift his body towards me, his handler took control of him immediately by yelling at him and intimidating him with the vicious bull hook. And this majestic animal began to sway his body in distress. He was shackled by unreasonably short chains—one of his hind legs was tethered to a cement pole and his front legs were chained like a handcuffed prisoner. He had no wiggle room and was forced to stand in one spot. This made it extremely difficult for him to balance the weight of his massive body.

Later that same month, Ayyappan would participate in the Thrissur Pooram celebrations. I vividly remember how my cameraman and I were busy filming the massive crowd that was dancing and singing away. People seemed totally oblivious to the suffering of one hundred elephants that were lined up and shackled heavily so that millions of people could rejoice. Suddenly, we heard the sound of a pistol. The ignorant revellers dancing atop Ayyappan had shot confetti during the umbrella ceremony and spooked the animal. We captured the dramatic footage of Ayyappan turning around and trying to bolt as his handlers implemented tremendous force to bring him under control.

The next time I visited Ayyappan was in November 2014, when he was in his musth cycle. During this period, the testosterone and energy levels surge and the bulls are overwhelmed by the urge to mate. In the wild, they wander for hours on end and deplete their energies by mating and fighting with other bulls. But in captivity, Ayyappan's primal instincts were cruelly denied, and the shackles were tightened severely. He was desperately trying to break free and was so frustrated that he began tossing palm branches and rocks at the mahouts.

During that visit, I had learnt that in order to weaken him, Ayyappan was often deprived of food and water during his musth. It seemed unfair that the handlers were feeding and bathing the other two elephants in the same compound, while ignoring Ayyappan. I dared to pick up the hose and began hosing water at him from a safe distance. I will never forget how Ayyappan desperately opened his mouth and drank water continuously for nearly fifteen minutes. That's how thirsty he was! Haunted by the atrocities that I'd seen, as soon as I returned to Bangalore that evening, I called his owner, Sundar Menon, and reported what I'd witnessed. He assured me that the mahouts were being monitored by a CCTV camera that was installed around all three elephants in his compound. But he fell short of answering why Ayyappan was so badly neglected.

In the ensuing weeks and months, I frequently phoned Menon, inquiring about the three elephants in his compound—Shiva Sundar, Ayyappan, and my beloved, Lakshmi. One day, he casually told me that Ayyappan had been allowed to mate for the first time. Apparently, they had unshackled only his front legs, and with his rear legs tethered to the usual spot, he mounted Lakshmi, who was chained right next to him. Unable to bear my distress, I ended the call abruptly.

What a heinous crime against nature! Menon had just told me that Lakshmi had been raped by Ayyappan! Although it may sound preposterous, there's no better way to describe what happened to her. Lakshmi must have been traumatized. I was later informed that she tried to escape, but defenceless and chained, she was unable to move. She could not save herself and was forced to mate.

When a woman is forced into sex, we call it 'rape' and the convict is severely penalized. Although in this case, Ayyappan is not the convict, but rather the people who facilitated this heinous crime should have faced hefty monetary fines and a jail sentence. So, why should it be any different if an elephant, or any animal for that matter, is forced to mate? Have we forgotten that humans are also part of the animal kingdom?

Rumours of Lakshmi's pregnancy spread like wildfire. This was of grave concern as I had read of captive elephants killing their young ones, not knowing how to nurture them. In the wild, females of the herd rear the baby, very much like in human families. But Lakshmi had never seen a herd. Even if birthing was successful and Lakshmi embraced her calf, the baby would no doubt be at the mercy of ruthless human beings. People would torture and train and then exploit the innocent calf to make money. Fortunately, after a few nail-biting weeks, the medical results turned out to be negative and Lakshmi wasn't pregnant after all.

This episode evoked my own childhood nightmares. At ten years old, once when I'd been playing hide and seek with my friends in a courtyard, a predator lured me into his home. He then locked the door, stripped himself, and tried to fondle me. I yelled and cried for help, but he told me to shut up and threatened to hurt me. I kept dodging him and ran onto his verandah. Luckily, I found the door, flung it open and dashed out. My grandma heard me sobbing and rushed to me. I threw myself into her arms, tears flowing endlessly. I told her what had happened, and she tried to console me. She also made me vow that it would be a 'secret' between us and that I would never repeat the story.

My beloved grandma died in 1995, and, I think, after all these years, she would be okay with me shedding light on the harsh realities that everyone suffers by inflicting pain on others. The agony that Lakshmi and I have been through is no different from what vulnerable girls or boys or women or other animals suffer. By breaking my silence, I hope to help people realize that they are not alone. Many of us have experienced similar episodes and we can heal ourselves first by sharing and releasing our suffering, and in so doing, we can also aid others with their healing journey. We are all one, therefore, if one suffers, everyone suffers.

When I came of age, some of the most trusted and close male relatives treated me like a commodity that they could use any way they wanted. In particular, I vividly remember an occasion when my mother and I were travelling in a packed train with one of our relatives. I was holding onto the handle dangling from the roof, trying to balance myself. Suddenly, I felt a strong grip on my right breast. I turned around to discover that this relative had cupped his palm inappropriately. As I gave him an angry look, he glanced away, shrugging off my inability to defend myself in the packed carriage of the train. Apparently, this slimy man had a tendency to behave in this matter with every woman he encountered. His wife confessed to my mother that he would use any excuse to touch a woman.

Harassing women and treating them like a commodity has been a social norm in India since time immemorial. The 2012 case of a young woman who was gang-raped on a private bus and her body tossed onto the streets of New Delhi with an iron rod having been thrust into her vagina, created a global outrage. Her partner was also beaten up and thrown off the bus as he cried helplessly all the while.

The girl was transported to Singapore, where she eventually died. But the rapists' lawyers argued that it was the girl's fault. She should not have been on the streets so late at night. In fact, it was only 8:00 p.m. But don't you think women should feel safe to walk wherever they want at any time of the day or night? Why should a girl not have the same right as a boy? Although there were countrywide protests to protect women and girls, little seems to have changed as abuse and rapes continue unabated.

If women are treated in this manner, why does it surprise anyone that so many animals and Mother Nature herself are being constantly exploited? How can we expect elephants to be treated with compassion in a country that condones the abuse and devaluation of women? It is indeed a vicious cycle of abuse and subjugation of every vulnerable sentient being in a patriarchal society where masculine energies are destroying the threads that make up the very fabric of humanity.

The parallels between the mistreatment of women and elephants in India cannot be ignored. Women are exploited for dowry—in the form of money, goods, or an estate that a wife brings to her husband at marriage. And elephants are likewise force bred to quell man's insatiable lust for material wealth. Both of these circumstances are caused by greed, a lack of empathy, and a feeling of separation. This all too sadly manifests as social and spiritual upheaval in our world. Unless and until women and girls are cherished and respected, there is little hope for animals and the natural world. And this will ultimately cause the implosion of human civilization itself. Remaining silent on these injustices will only perpetuate them. In fact, those who turn a blind eye, or don't speak out, are actually facilitating these heinous crimes. It's about time we stopped hurting each other and our precious non-humans!

In India, even today, despite all the advancements, daughters are considered an economic liability, and families are driven to destroy female foetuses. Female foeticide is the deliberate act of aborting a foetus, solely because it is female. This crime against the females has insidiously crept into the society over the past four decades, ripping apart the rich social and cultural tapestry of India. Millions of girls have gone missing in the past thirty to forty years, with India's 2011 census showing a serious decline in the number of girls under the age of seven. It is estimated that eight million female foetuses may have been aborted in the past decade. The organization Global Girl Power claims that in the last century as many as fifty million girls have died in India because of female foeticide.

This began when ultrasound technology arrived. Medical professionals discovered a goldmine and quickly exploited and misused the technology. They destroyed female foetuses at an unprecedented rate, making sex selection a lucrative business. An infamous ad in the 1980s offered the chance to save a potential daughter's dowry payment when she married. The ad read, 'Pay 5,000 rupees today (to have an abortion) and save 50,000 rupees in dowry payments tomorrow.' Such reckless actions have caused a significant gender disparity in India, with only 943 women for every 1,000 men as of March 2015. This is a huge difference in a country with a populace of 1.38 billion people as of June 2020. Indeed, India is now sitting on a ticking demographic time bomb with 37 million more men than women. Most of these women are of marriageable age, given the relatively young population. The cascading social crisis in the north-eastern parts of India is simply unimaginable; young girls are bought from their parents and sold in the marketplace.

Something similar is unfolding in elephant societies as well. Male elephants are killed for ivory and captured illegally for cultural festivals in India, which is causing a serious decline of male elephants in the wild. During one of my interviews with elephant expert Dr Raman Sukumar, he said that in some parts of India, the male-female sex ratio in the elephant population is as high as 1:60. The exploitation of elephants for profit and women for dowry are woven together by common threads—cultural myths driven by materialism and power. Turning a blind eye to the situation is causing social dysfunction in both human and elephant societies.

The irony is glaring! Although people worship Lakshmi, Durga, Saraswathi, and other goddesses in India, women are still raped and abused. Elephants are considered to be the embodiment of Lord Ganesh, but they are tortured and enslaved. Tigers are considered to be a vehicle of the Goddess Durga, and yet these regal animals are being poached to extinction.

To return to our narrative about the bull elephant, Ayyappan, fast forward to February 2016. The first thing I did when I landed in Thrissur, even before checking into my hotel, was to buy fruits for the elephants I was about to meet. Then I went straight from the airport to Sundar Menon's mansion. I finished feeding the two elephants—Shiva Sundar and Lakshmi—as she had distracted me away from Ayyappan. But just as I was approaching this young bull to feed him pineapple, an infuriated man came dashing towards me, yelling and screaming, telling me to get out of the complex. He was the handlers' supervisor. I had never seen this side of him during the past few times I'd been there. In the past, he seemed kind and compassionate, always greeting me with warm and welcoming smiles. However, this time he was nasty.

I later learnt that the temple authorities had ordered the supervisor to prevent me from seeing or interacting with the elephants. The last thing I remember as I was being escorted out is Ayyappan stretching his trunk for the pineapple that he thought was forthcoming. Only if I had gotten to Ayyappan a bit earlier, I could have fed him. To this day, I struggle with guilt for not having fed him what was meant for him—his share of the fruits. A year went by, but despite repeated attempts, the owners and temples denied me access to these elephants. They were furious that the truth had been revealed. And then . . . a bombshell was dropped on me!

On 17 December 2016, I received a phone call early in the morning from the vocal animal welfare activist Venkitachalam. He broke the devastating news—Ayyappan had died that day. The post-mortem results revealed that his intestine was clogged with date seeds. Elephants are only meant to be fed seedless dates. Notwithstanding the physical and emotional suffering perpetuated by his tight shackles, it is hard to comprehend the trauma that Ayyappan must have gone through in his final hours, added by the agonizing stomach pains. Humans can visit a doctor and express their suffering, but Ayyappan suffered silently and eventually died. The only relief is, at least, he would suffer no more, finally free of shackles, of torture, of deprivation.

This majestic animal had been kidnapped from the wild, ripped apart from his family, and tortured physically and psychologically. Had he been left in the wild, he would still be wandering freely, mating, and producing beautiful babies. But he was captured, emasculated, enslaved, and brutalized—all at the hands of the mercenary men. Later in the day, I received an e-mail containing photos of Ayyappan's body covered in garlands and flowers. An oil lamp had been lit near his head—a scene all too familiar. Ayyappan was being revered in death, after having been tortured all his life. Here again is this incredibly tragic paradox! You can see what these poor creatures are up against, on virtually every front.

Ayyappan is just one sad example of the ultimate injustice that these elephants face. His cohort Lakshmi at the Thrissur compound is another. And then there was Ramabadhran, the elephant with the paralyzed trunk who had died. Shiva Sundar, the superstar of Thrissur Pooram also met the same fate. Apparently, his entire digestive system was clogged up and infected. As I write this chapter on 28 May 2018, seventeen elephants have died in Kerala in less than five months into 2018. But the elephant keepers continue to do the same things over and over again, feeding the same type of food, parading the elephants in multiple festivals every day, and neglecting their basic welfare. If they can't care for the animals that fetch them so much money, then why bother keeping them?

Here's the harsh reality that people of Kerala simply will continue to deny. Elephants belong in the wild, not in a temple or someone's backyard. They are enormous animals and can be controlled only by brute force. No matter how gentle and amazing they may be, they are also unpredictable. Even their slightest, most innocent attempt to express their love could set in motion a chain of unanticipated, tragic outcomes. So, why keep elephants in captivity at all?

At least, one elephant was lucky enough to be released from the insidious clutches of humans. His story went viral on social media, drawing world-renowned celebrities. He is none other than Sundar, the poster child of all temple elephants. This is the same bull elephant shackled in that infamous Jyotiba Temple in Kolhapur, forced to stand on the fuchsia-pink floor. But I hadn't yet seen this brave warrior after he had been relocated to his new home at the Bannerghatta Biological Park (BBP). So, it was with great anticipation that I set off to see him . . . my final trip in India before returning to Toronto.

When I reached the BBP, Sundar was nowhere to be seen. He had wandered off into the woods to enjoy his newfound freedom. It took us more than an hour to find him. When we finally did, he was deeply engrossed in grazing, his backside facing us. But feeling our presence, he suddenly turned around. My heart skipped a beat when I saw him, for he was no longer clad in shackles. He was relishing his freedom. His mahouts gently cajoled him out of the woods. They used no weapons, instead, they just kindly patted his thighs to urge him along. Reluctantly, he left the forest area and walked alongside his minders, who used a reward-based positive reinforcement method to follow their commands. Sundar knew all too well that he would be treated with goodies for listening to them.

After having witnessed the hellhole in Kolhapur where Sundar had been tortured by his ruthless mahouts, it was a great relief to witness this kind of love bestowed upon him by his new mahouts. Not only had he adapted well to his new home but also had made several friends, especially a young female who was now his constant companion. I was informed that just when he had begun courting this female elephant, Vanaraj, the bull whom I had seen during my first visit, attacked Sundar, injuring his belly. He was given a rigorous course of antibiotics and was healing well. Sundar was also receiving nutritious meals, including protein-rich, horse-gram lentils, rice, coconut, turmeric, and watermelon, as well as other delicious fruits. A large enclosure where he would remain isolated during his musth cycle had also been constructed.

When I approached him, he stuck his trunk out of the enclosure and sniffed me out with a sense of familiarity. He then stuck his leg out, too, so he would receive some treats. This is how the veterinarians provided foot treatment, and then rewarded him for his co-operation. I had taken some bananas and fed him the fruit, then tickled the bottom of his foot affectionately. Today, Sundar socializes with seventeen elephants, including a baby, in the forest camp where wild elephants frequently visit and mate with the camp's females.

This trip to India had been a whirlwind, but it was very productive. I returned to Toronto armed with my newly shot dramatic footage, interviews, and most importantly, the climax that had transpired with Lakshmi. The following day, I was supposed to meet with Digby and Rob to discuss the change in direction of the film. However, Rob phoned me on the day of the meeting, sounding a bit unsettled. I instantly sensed something amiss. Just when I'd thought things were flowing smoothly, here I was, faced with a new challenge, and off on another roller-coaster ride!

Chapter 15

The Valleys and the Peaks

Rob broke the earth-shattering news—he had been offered a full-time job and was intent on accepting it. Thus, he was quitting. This time around, we chose not to advertise for a video editor. Instead, Digby made some inquiries and through a friend of his, we received three recommendations. After conducting phone interviews, we zoomed in on Billy, given that he seemed to have plenty of time at his disposal. So, we interviewed him in person.

Over six feet tall and lean with a clean-shaven head and blue eyes, Billy walked into my home office. Digby and I asked him several technical questions to ensure that he could work with the programme that our previous editor had used. Billy confidently said he could. I clearly explained that my budget was limited and told him that I couldn't afford to rent out an editing suite in a production house. I also reiterated that we would have to edit either at his place or my home office, and he agreed to everything we asked.

Now, before I proceed any further with this chapter, I'd like to make a disclaimer. I'm not trying to blame or criticize anyone, but I feel it's important to share the overwhelming personality challenges that I was faced with during the final stages of production. As a female, directing men in a male-dominant industry, it was unbelievably difficult to get this man, in particular, to follow my instructions and do what was needed to make *Gods in Shackles* impactful. But by accepting the circumstances and communicating openly, while also seeking some guidance, a few earth angels and a higher power helped me shatter the glass ceiling.

Back in my home office, I began to gasp for breath. I am incredibly sensitive to cigarettes as they aggravate my mild asthma. And Billy's shirt emanated a potent odour of cigarette smoke. In fact, he couldn't complete a sentence without the smoker's cough. But despite my apprehensions, in my desperation to get the movie out into the world and expose the dark truth behind the glamorous cultural festivals, I hired him anyway.

The next day, when he picked up the hard drives, I was quite impressed by his proactive approach. However, he called me the following day with some bad news. His system kept crashing as the content was too large for it to handle. In other words, his equipment was inadequate for my massive project. Ignoring the second alarm bell, I asked him to work through the problem, and by the end of the week, he had it fixed.

We had initially agreed to edit in my home office. However, he phoned me the next day with another tentative news. His wife wanted us to edit at his house. So, the next day I arrived at their place where an adorable little puppy greeted me with warm hugs. I couldn't help but pick him up. He licked my face and showered pure love on a complete stranger. But apparently, his wife felt 'jealous' as Billy put it. This meant that we couldn't edit at his place, nor could we edit at mine. The only way out of this dilemma was to secure an editing suite. But I reminded him that I had no budget for this.

The challenges seemed insurmountable. I was still meditating daily and repeating my intention religiously. In despair, I called Digby, my scriptwriter, and Paul, my media adviser, hoping that they could help me figure out a solution or a compromise. I suddenly realized that Billy hadn't signed his contract! So, contractually his actions weren't putting him in breach of anything that I could legally take him to task about.

Luckily, Digby's friend, Howie, owned a film production house in the heart of downtown Toronto. And due to a slow period, Howie had a vacant editing room in his facility. Using his considerable charm, Digby persuaded his friend to let us use the room but operate our own equipment.

The next day, I visited Howie to thank him. Meeting him for the first time, he greeted me with a warm hug. Approximately fifty-five years old with a distinguished broad forehead and glasses, Howie was blatantly silly. He had absolutely no inhibitions, and the F-word was an integral part of his vocabulary. In the twenty minutes I spoke with him, I realized that Howie was a genuine and sincere person. I felt I could trust him. And since he had recommended Billy, I was now convinced that Billy must also be a good person whom I could trust to get the job done.

In mid-July, Billy brought along his equipment and we began editing at the post-production house. He did a great job sectioning the entire movie into themes and my transition to this third editor felt seamless. Perhaps this was too good to be true! Indeed, within just three days, Billy's computer began to crash almost every five minutes. With nine hard drives plugged into his hardware, it couldn't handle the load. Soon, it became clear that Billy's editing system was ill-equipped for my massive project. Less than a week into editing, additional technical issues began to creep up.

Billy became frustrated and communication between us became quite fraught. I couldn't ask him any questions or offer suggestions as he erupted like a volcano when I did so. He became aggressive and abusive. Meanwhile, the stench of cigarettes in his breath and clothes started giving me headaches. For the longest time, Billy kept blaming the footage we had shot using different cameras. However, soon the truth had surfaced—his system was clearly at fault. It took him two weeks of wreaking chaos and confusion to concede that his editing system was inadequate. I also realized that contrary to what he had claimed during the interview, he was unfamiliar with the editing software that Rob had used prior to bringing Billy on board.

But here was the greatest issue: I couldn't technically hold him accountable because he still hadn't signed his contract. He was charging me by the hour and my crowdsourcing funds were shrinking. Given that not many editors were ready to work within my shoestring budget, it was hard to let him go. All this added more tension to the already tense atmosphere. A strained meeting headed by Howie ensued. However, despite the fact that Billy had been dishonest right from the beginning, I compromised. I watered down his contract so that he would finally agree to its terms and sign it. But the working conditions never changed. Billy and I became resentful of each other, and this began to impact the quality of the film itself.

I turned to my very first editor, Maneesh, for help. He not only generously offered up his editing suite but even helped to transfer the content. Billy and I then resumed our editing, picking up where we had left. The occasional computer crash was natural for a program the size of my film. But instead of being happy, Billy became more defensive and resentful that he had been proven wrong. Instead of letting go, he was clinging onto his belief and kept justifying that his system was not the problem. We were so close to completion, and yet so far away.

Finally, when we were within spitting distance of completing the edit, I called it quits. It was too much to handle. I don't know why I hadn't sought Paul's counsel before things got out of control. However, that evening, Paul talked me through the situation, and helped me realize that I was so focused on ending the abuse of the elephants that I had been neglecting the emotional toll that Billy's abuse and aggression were taking on me. It took a compassionate friend and mentor like Paul to help me realize that I should never have tolerated Billy's abuse. I should not have hired him in the first place, especially after my internal alarm bells had gone off loud and clear at least three times. This taught me to never ignore my inner voice and put myself in harm's way again.

By now, my funds were running very low, and I had exhausted my entire budget for editors. We still had fifteen minutes of the film to edit, motion graphics to create, and sound mixing pending. I had also invested all of my personal savings in addition to the donations I'd received. And I realized that I had under-budgeted the post-production phase, having failed to raise contingency funds. But my supporters rose to the occasion, once again.

Monica Kelley, a teacher and a key force behind our initial crowdsourcing campaign, had organized a 'Global Walk for Temple Elephants' for that initial campaign. She and her family lived in Geneva, New York, where she organized another global walk for the elephants. As a part of that, she invited me to her school in October 2015. It was delightful to finally meet her in person. She was a true soulmate in the work we were doing together on behalf of the elephants. It was a great honour to visit the Geneva High School where students had plastered the walls with welcome signs. Teachers, heads of departments, and the principal had organized a gathering for me with the students in a packed library.

Later, I also met Monica's students, but the most touching experience was meeting her family. That evening, her parents and in-laws joined us for dinner at her place. Her adorable son and daughter entertained us with dancing and games, as her three dogs hung out with us. She raised more than twenty-five hundred dollars and created big ripples of success for our campaign. Following her success, other elephant lovers launched a matching-donations campaign. Amanda Loke of Hong Kong, who had already donated a significant amount of money for the film, contacted me saying that she wanted to make another substantial contribution. This helped me to supply the much-needed funds to hire what would be my fourth film editor.

Now that I had the funds, I turned to Howie, Digby's colleague, once again. We had an open and honest conversation about the current situation. Howie then connected me with a gentle soul named Shaun, who lived two hours outside of Toronto. Shaun and I met the following week. As he entered my condo, he carried a calm and peaceful presence. His piercing blue eyes scanned my living room, seemingly trying to figure out a suitable spot to edit. He smiled frequently and spoke gently, which was reassuring to my tender heart that had been traumatized. Then, Shaun pulled out his laptop and confessed that his system was inadequate for such a massive project as *Gods in Shackles*. But fortunately, Maneesh's system was sitting in the corner of my living room. At this point, the pieces of the puzzle fell into perfect place. I hired Shaun, and our final push to the finish line began in earnest.

The only issue now was his travel time. He had to drive a long way to get to my place in the morning and that would cut into our day. Also, driving four hours a day was an exhausting proposition. After a couple of days of commuting back and forth, his wife and Shaun felt it would be best for him to stay in Toronto. But the question was where? He stayed with his friends for a couple of nights but that didn't work out. I then offered him a sleeping bag to sleep on my couch, opening my home to a total stranger.

Late-night edits were common—most days we edited for twelve to fourteen hours. The problem with such an arrangement was privacy. We were in each other's immediate presence day in and day out, which made us impatient and agitated. So, once again—and yes, you might have guessed it—despite Shaun's calm temperament, we had a major meltdown. With just three minutes left to complete the film, we got very close to calling it off. But fortunately, we talked things through, and cooler heads prevailed in the end. On 30 August 2015, we completed the final cut.

It was nothing short of a miracle that Maneesh had offered up his editing system for our use at the perfect time. Without it, I don't know what I would have done. I remain extremely grateful to Shaun for the wonderful contribution he made to *Gods in Shackles*. This was another example of the universe providing when I needed it most.

After Shaun completed his job, Maneesh stepped back into the picture to give it the finishing touches—colour correcting, image sharpening, stabilizing the shaky videos, and making other key technical fixes. As fate would have it, Maneesh would become an integral part of the *Gods in Shackles* documentary. He also ended up designing our motion graphics. This was very critical in portraying the brutality and barbaric torture involved in capturing baby elephants from the wild and training them. They say that when you let go, what is meant to be yours will return to you and I suppose that Maneesh, in this instance, was a good example of this.

We were now in the final phase of post-production, and I still needed a sound mixer. Paul introduced me to Tom Mullins, with whom I made a wonderful connection. His calm demeanour during the sound-mixing process was necessary to put out the best. Through synchronicities, he came on board at a critical juncture in the film's evolution.

On 21 May 2016, the entire production of *Gods in Shackles* was complete, in perfect timing that had been ordained by the cosmos. Sure, I had done my bit, but my part was really to follow and not to lead. I fervently believe that God, a higher power, or the universe, or whatever you want to call it, was in charge of this project, overseeing it at every turn.

Chapter 16

The Movie Launch

Before the film release in India, we had to obtain necessary approvals from the Indian government. The first challenge in this phase was to have the Animal Welfare Board of India formally approve *Gods in Shackles* and provide a 'no objection certificate' (NOC). This, in part, was necessary to establish that elephants had not been forced to perform for the film. This was a fairly simple and easy process, and fortunately, the NOC was secured in a matter of four weeks.

The next step was to obtain the green light from India's Central Board of Film Certification (CBFC), a statutory censorship body that falls under the Ministry of Information and Broadcasting. CBFC India is considered to be one of the most powerful film censor boards in the world due to its stringent regulations. Without the board's censorship certificate, films cannot be publicly exhibited in India. But the process to apply for the censorship certificate would be anything but simple. I was also uncertain about what the certification process would entail as I had never produced a movie for release in India. One of the animal welfare activists had promised to assist. Yet, he ignored my numerous follow-up e-mails.

Left to my own devices, I surfed the net and found the online application. But it looked complicated, and the terminologies were hard to comprehend. Still, after reading the application numerous times, I realized that two short segments of the movie could potentially pose a problem—one was a particularly gruesome scene, exposing the brutal beating of a bull elephant after he had emerged from his musth, and the second one was a culturally sensitive comment by one of the participants. Both of these had the potential to spark public outrage.

On a chilly winter night, at 12:00 a.m. in Toronto and 10:30 a.m. in India, I mustered up the courage to phone the CBFC's mainline. It was the Bangalore censorship wing. I felt the fear but did it anyway after offering a quick prayer, as I'd been doing throughout this journey. In order to be as thorough and accurate as possible, I had constructed a list of questions so that I could have a clear understanding of the process. The requisite application had to be flawless, leaving no room for rejection.

When I phoned the CBFC office, a pleasant male voice answered. His name was Babu, and he sounded like a Keralite. At first, this scared me even more, for the movie was exposing the atrocities in his home state. If he was among those screening the film, I could be guaranteed that the CBFC's stamp of approval would not be forthcoming. But despite my serious doubts, I spoke with him in broken Malayalam, and fortunately, he turned out to be very helpful and personable. Babu patiently guided me through the step-by-step process, articulating clearly how I could expedite the CBFC's approval. He then said that the applications had to be hand-delivered. But I was in Canada and this would be impossible. So, I was now tasked with appointing someone in Bangalore and authorizing that person to obtain the censorship permit.

But first, I would have to create a list of every shot in the movie for the CFBC. Now, through the entire production and post-production phases, I had probably reviewed the footage well over a hundred times. And just when I thought I could take a break from the traumatic images of torture and abuse in the film, I was put through another resilience test. In an ideal world, the producer's assistant would create the shot list, but I had no assistant. So, I had to do it myself. The tasks ahead were too overwhelming for one person to handle. Regardless, I made a schedule of 'things-to-do' and set out to cross them off my list—one by one. Within one week, I had managed to compile all the documents necessary for the censorship bureau, including the dreaded shot list and had also made requisite copies of DVDs and Blu-ray discs.

The next step was to find someone living in Bangalore who could act on my behalf, and whom I could trust. I instantly knew the person who I could prevail upon. I phoned Suparna and she promptly agreed to help out. After receiving my courier, she went above and beyond the call of duty by driving as many times as necessary to the other corner of the city to deliver the application documents. Suparna is a perfect example of how the right people were placed in my path at critical junctures to keep moving the project to the next level.

This happened so frequently that I almost began to expect it. Of course, this required a huge amount of trust on my part and that was the specific personal journey that I was on through the making of this movie. I had to learn to trust the universe and that everything was always working . . . for the highest good of people and elephants alike. My convictions became even stronger after the film received the CFBC's approval. Now, I could start planning the screenings of *Gods in Shackles* across India, which eventually got scheduled between May and June of 2016.

Meanwhile, in the last quarter of 2015, well before the completion of the final cut, I had begun submitting *Gods in Shackles* to several film festivals. Lo and behold, on the very first day of 2016, I was informed that the movie had won the Hollywood International Film Festival award! This would be followed by eleven more awards in the ensuing months. At least half a dozen of them, including the prestigious

Cayman Island Film Festival Award and the Power Brand Glam Bollywood Award selected *Gods in Shackles* in the category of Best Documentary. But the greatest surprise and honour was the nomination at the United Nations (UN) by the International Elephant Film Festival (IEFF). This is a joint venture of the Convention on International Trade of Endangered Species (CITES) and the Jackson Hole Film Festival. The public acknowledgement of this nomination would take place during the inaugural festival at the UN headquarters in New York on World Wildlife Day (3 March).

The way this came about is nothing short of a miracle. When our cinematographer, Tony, had initially told me about this festival, the film was still incomplete. And I had assumed that the festival would not accept unfinished projects. But, to my surprise, they accepted our rough cut. Furthermore, I didn't think that *Gods in Shackles* would even make it into the festival, let alone get nominated at the UN.

One of the things that I was becoming aware of was—no matter how many awards the movie received, feelings of 'not good enough' kept hounding me. It had to take the UN nomination to break through the mantle of my inadequacies and affirm my worthiness. It instilled a sense of trust in my innate potential. I was now ready to travel to New York for this poignant day.

It was a chilly winter morning in Toronto on 3 March 2016, at approximately 3:00 a.m. when I was driving to the Pearson International Airport to catch my flight to New York. I had stayed up all night because, for some reason, I was paranoid that I'd miss the airport exit on the highway—exit 409—and, thus, miss the 7:00 p.m. flight. Talk about a self-fulfilling prophecy. I did miss the airport exit, which I realized only after I had overshot the next two exits. I then took the third one and drove back like a maniac on the eastbound highway. It was another one of those close calls. I almost missed my flight. Fortunately, since I had only a carry-on suitcase, I was able to board my flight on time.

Upon arrival in New York, I took a cab to the UN headquarters. Once on the street, standing across from the institution, I glanced up at the tall building. A crisp winter breeze stung my senses as I choked back tears. I was astounded that I would be entering this iconic building I had seen on the television so many times in the past.

I looked at the large clock on one of the buildings and realized that I had enough time for a cup of coffee. So, I walked up the street hoping to find a coffee shop. But there were no Starbucks or Tim Hortons in the vicinity. Instead, I found a Subway sandwich shop after walking an additional two blocks through the bone-chilling wind. Once inside, frigid and in my desperation to thaw out, I clasped a hot cup of coffee in my palms and sat at the table farthest from the door.

I had invited Monica Kelley, the biology teacher from Geneva High School, to the UN event. She had been one of my primary fundraisers for the film, thereby earning the title of honorary associate producer. Because she lived in New York, it was fitting to invite her to the UN ceremony. Sipping my morning java, I called Monica to find out her estimated time of arrival. As it turned out, she was just a block away on 45th Street, also drinking coffee, and she showed up within five minutes.

It felt like a reunion of long-lost friends as we hugged each other. It was only in October 2015 that Monica and I had first met, but within a short while, we had developed a deep bond. Monica then explained that she had taken the day off from school and boarded an early morning bus from Saratoga Springs just to be present at the event. She had to return home the same day. After catching up on life, we made our way back to the UN headquarters where there was a long line-up. But when it was our turn, we were promptly blocked. They would not allow me to carry my suitcase for security reasons. The guard directed us to a hotel across from the UN building where there was a short-term storage space. So, we walked back, racing against the winter wind again. To our disappointment, they rejected our request. I now had to think on my feet.

Suddenly, the same Subway sandwich shop, where I'd had coffee just hours back, and the friendly Indian waitress came to mind. Monica and I hurried back there together, and despite the morning rush, the kind woman listened to my pitiful story and agreed to stow away my tiny suitcase in the kitchen. This Indian woman who worked at Subway—had been placed in my path to ensure a smooth flowing of events!

By the time Monica and I walked back to the UN building, we had lost about thirty minutes. However, there was still enough time to stroll through the halls and take photos. It felt surreal entering the magnificent auditorium where global decisions are routinely made. This is where the award ceremony was about to take place. Dignitaries from around the world were arriving and being escorted to their seats.

I glanced around and realized that I was seated right next to Kristin Davis, a Hollywood actress, who played Charlotte York Goldenblatt in the HBO romantic comedy series *Sex and the City*. Kristin has been quite an activist on behalf of the elephants, and I had to pinch myself to believe that I was one of the few women sitting with her in the front row.

After the welcome speeches and ushering in of the film festival personnel, the winners were announced. *Gods in Shackles* did not win the Asian elephant category, but to sit amid like-minded people and discuss elephant conservation was a great honour. Soon after the ceremony, the CITES Secretariat, John Scanlon, with whom I had communicated a few times, walked down the stage and congratulated me. I had seen him on television but to spend an evening with him and other decision-makers at these red-carpet movie screenings was beyond what I'd ever expected.

When I'd embarked on my mission to expose the atrocities against the elephants, I could have never imagined that the journey would lead me to the assembly chamber of the United Nations, nor the film being nominated at the UN. But to me, the greater significance of this nomination was that it offered a pulpit that I could use to voice the dire need to protect Asian elephants and create awareness about their plight. My journey to the UN, despite a tumultuous start—from missing the airport exit in Toronto to being denied access to the UN itself—had worked out in the end.

As soon as I returned home to Canada, I called Suparna to find out if she had any news from the CFBC. And as it happened, they had contacted her that very afternoon, revealing that the film had received a 'U' (Unrestricted) certification, which would allow everyone, including children, to watch it—after one sensitive scene had been removed. To remove this bit from the film would not be easy. It would entail dismantling all the tracks—including the audio, video, graphics, motion graphics, and titles—and require numerous hours of editing. As vital as this was, I had exhausted my budget. So, I had to return to the drawing board. I contacted Maneesh and explained what the CFBC needed. As always, he was ready and willing to help. But I had to pay him at least the minimum wage.

Once again, I was filled with doubt and feared approaching my supporters for more donations. This need for funding seemed like a vicious, never-ending cycle. Every time I thought I was done asking for money, something else happened, and I was pushed out of my comfort zone to do whatever was needed.

What kept me going were the nightmarish scenes of elephant torture that I dreaded to watch. After having witnessed so many atrocities, there was no turning back. Elephants had touched me profoundly, giving me a sense of purpose and conviction to deal with the insurmountable challenges that my endeavour involved.

And like before, one way or the other, I somehow managed to achieve my goals. I raised some funds, edited out the scene, couriered the DVD and relevant materials, and by the end of March 2016, I received the stamp of approval from the CFBC. This was indeed a huge relief as it had been weighing our entire team down for more than three months. Now, I had to find a way to release the film in India, which required finding a distributor for it.

Over the next few weeks, I approached several distributors but given the culturally sensitive nature of the documentary, this would prove to be problematic. I also contacted India's television networks, but none of them responded to my e-mails. So, now, I was left to disseminate a multiple award-winning and UN-nominated film on my own. I had already invested a lot of money from my own personal savings in addition to raising more than $US140,000. But after spending all that money, time, and energy, it felt like my efforts would be for nothing. I was feeling crushed morally, emotionally, and spiritually.

Just when I was losing hope, a reporter from the National Geographic network e-mailed me, wanting to do a story on the plight of elephants. I was elated by the much-needed exposure for my film. My initial responses to noted journalist Christina Russo's questions were mechanical and did not conjure up too many feelings. In fact, they seemed almost irrelevant to my soul. But during the two-hour-long interview, Christina drew out my heartfelt emotions, allowing for a cathartic and healing exchange between us. Her questions allowed me to shed light on my personal journey, which really touched the reporter at her core. Christina then recommended another journalist to me so that the film could get even more exposure—a journalist from the *London Telegraph* who had been reporting on festival elephants.

The wave of popularity gained momentum. As a result, and by way of the most unexpected sources, *Gods in Shackles* received an endorsement from the revered conservationist, Dr Jane Goodall. She has been my idol since my university days when I pursued studies in biology. Her research on chimpanzees, revealing that they had feelings, had been initially rejected by the reductionist scientific community. But she never gave up. It took her decades, but her unwavering commitment to these closest relatives of ours brought acclamation to her labour of love.

Dr Jane Goodall also supported elephant conservation, and I had tried to contact her through my friend Paul Lewis. However, despite repeated efforts, he was unable to gain access to her. Then I suddenly thought of a powerful man in the UN cultural wing, who was introduced to me by a mutual acquaintance sometime prior. They wanted to potentially screen the film in various cultural and religious settings. He had casually mentioned that Dr Jane Goodall was a close friend of his.

A year had gone by without much contact with him. And at first, I hesitated to ask him for any help, fearing that I might appear opportunistic. Finally, I decided that, at most, he would reject my request. But as soon as he read my e-mail, without any hesitation, he connected me with Dr Jane Goodall's top aid and confidant. Within two weeks, Dr Jane Goodall screened the film online and afforded me a shining endorsement of it. More endorsements poured in for the film, far too many to list. Suffice to say, with so much media exposure and prominent endorsements, *Gods in Shackles* gained the much-needed momentum.

But I still could not find a suitable distributor. For my supporters, time and patience were running out. They wanted to see the result of their contributions. I was under great duress and, once again, I turned to my own devices. Every step of the way, I had been keeping my supporters informed of the film's progress. One day, I openly expressed to them the challenges I was facing in moving forward without a distributor.

Outreach is critical to changing attitudes and propelling conscious action, including policy changes. I had to raise funds to screen the film on my own in India in order to cover the cost of theatre rentals, equipment, transportation, and other related expenditures. I knew now that the film needed all the exposure it could get, which could perhaps attract a distributor. Thus, did the world, my world anyway, unite in a common cause once more.

Screenings began in North America with two passionate animal welfare activists offering to host benefit screenings in their hometowns. They were Dr Amy Shroff, a veterinarian, and Maureen Mahon, an interior designer. They rented a theatre and did a lot of advance publicity to ensure a good turnout. Because of their efforts, the world premiere of sorts kicked off in Boston on 5 June 2016, World Environment Day. What a perfect way to celebrate our planet's most vulnerable keystone species!

Amy was passionate about all animals and given that she had Indian relatives, we also connected on a cultural level, sharing similar stories. Apparently, she used to run an animal clinic. However, at one point, she sold it to use the money to rehabilitate animals, mostly street dogs and cats.

The big day arrived—the launch of *Gods in Shackles*! I had brought along some silent auction items for the screening—T-shirts, posters, carvings, bags, and other things. As I lugged these items and stepped outside the hotel, heavens opened up, releasing a heavy downpour. But we made through the slushy roads to the Elephant Walk restaurant. Here, the screening was to take place that afternoon. Amy's husband, Howard, and Peggy, a family friend, greeted us warmly and led us to the basement.

A familiar face from Toronto—Donald Quan, my music composer—had arrived with another musician, Chris, from Boston. They were setting up musical instruments to do a practice run before the live performance. Soon, people began to arrive. Many of them were individuals I had connected with via Facebook. It was so wonderful to finally meet these fellow elephant lovers in person! With the silent auction, background music, drinks, chatter, and vegan dinner, the evening was off to a great start. The place was bustling with excitement. After about forty-five minutes, the musicians played a couple of instrumental renditions and then, I performed a couple of traditional Indian songs. Then it was showtime!

People were profoundly shocked by the atrocities they witnessed and were crying, with some offering to volunteer. We also ended up raising some funds, which would prove to be invaluable for screening the film across India. As I boarded my flight back to Toronto, I realized that my endeavour to expose the elephants' plight through the film was indeed a sacred communion of people for elephants who had made *Gods in Shackles* a reality. Everyone came together to embrace a common cause and do their part in saving the elephants.

The dress rehearsal of sorts in Boston allowed me to be better prepared for the upcoming screening in Los Angeles, the world's entertainment hub. On 18 June 2016, I was all set to leave for California, my heart soaring, and my mind filled with mixed emotions. This would be my first trip to Hollywood where legendary actors and actresses thrived. It was also where Maureen, the interior designer, lived. She had been a loyal supporter since the launch of the campaign. After having read my plea to help the elephants by creating awareness about them, she had sent me an e-mail, offering to host a red-carpet screening in LA—an irresistible offer.

19 June—the big day—had arrived! The day promised to be hot and humid, with temperatures expected to soar up to 100 degrees Fahrenheit, around 38 Celsius, that weekend. Overlooking the hotel's breakfast room was a concrete jungle nestled between green, man-made patches, nothing like I'd ever seen before. LA projected the image of a synthetic planet as seen in the movies, replete with tall skyscrapers interspersed between small homes with tiled roofs made of mud-red slate.

Suddenly, my phone rang. It was my music composer, Janal, who had arrived in LA two days earlier. I invited her for breakfast as she was close to the area. We hadn't seen each other for months, so it was a splendid reunion. We joked about my singing debut on Skype and reminisced about how far we had come. After Janal left, I took a stroll on the streets of LA before returning to my room, trying to fulfil my yearning to be a part of the wild and free California culture.

Janal returned at around 3:00 p.m. to pick me up for the screening. And then off we drove to the Harmony Gold Theatre where a team of twenty-some volunteers were busy organizing tables and arranging platters of vegan food. We went straight to the dressing room and began to practise for what would be our live musical performance. It was now past four o'clock, but the screening's host, Maureen, was still nowhere to be seen. However, another pleasant surprise soon eclipsed my distracted musings about Maureen's whereabouts.

A member of Hollywood royalty—the late Garry Marshall—had endorsed my film! Garry is the brother of the famous actress Penny Marshall but also a renowned writer, director, producer and actor in his own right. He is, perhaps, best known for directing and producing the smash hit *Pretty Woman*, starring Julia Roberts and Richard Gere. It was an incredible honour to receive an endorsement from this Hollywood royalty. This was made possible by Wedil David, a Hollywood actress who had starred in Garry's film, *Valentine's Day*. I connected with her through People for the Ethical Treatment of Animals (PETA), the renowned animal welfare organization that helped to promote the LA screening and invited Wedil, who, in turn, invited Garry Marshall. Although he was unable to attend, he produced a thoughtful thirty-second video in support of the film.

It was getting harder to wrap my mind around the abundant synchronicities that were continually unfolding, but that's what happens when you're truly in the flow of your life's work. At approximately 4:30 p.m., a tall woman with shoulder-length, dark brown hair and gorgeous grey eyes walked into the dressing room. She was wearing a grey knee-length sleeveless dress and silver jewellery. This was Maureen. She flashed a warm smile and informed us that the VIPs were waiting outside. However, I was still not ready as our song practice had delayed me. Nevertheless, I quickly donned my cream-coloured sari with its red border and my traditional Indian jewellery, slapped on some make-up, and then ran to greet everyone.

It felt surreal to be in LA on the red carpet with TV cameras waiting to interview me. Posters of the film had been plastered on the walls and in all the showcases. Photographers were snapping photos of the various guests against the backdrop of a massive *Gods in Shackles* poster. People were sipping wine and socializing, and everything was flowing smoothly. The anticipation was palpable as people were ushered into the auditorium and guided to the seats. It was 7:30 p.m. sharp. The celebrity emcee, Mark Thompson, a television and radio host, introduced my team. My music composers, Donald and Janal, took to the stage and tested the mics one last time. As I walked up on the stage, the computer wire got entangled in one of my feet and broke. We heard a chorus of 'ooohs' and 'awws'. Panic swept through the hall. Not a good omen, I thought. Our host made small talk to fill up space, but soon, Donald managed to get the system back up and running.

Janal then played the keyboard, as I closed my eyes and performed 'Vakratunda Mahakaaya', a song from the *Gods in Shackles* soundtrack that I had sung. Subsequently, Mark introduced me and the team, and we were greeted with thunderous applause. I gave a short speech, Garry Marshall's clip was played, and then I introduced the film. As the movie came on the screen, I could not have been happier. With perfect sound and impeccable quality, I watched my baby flawlessly come to life with incredible class and style.

There was utter silence even after the movie ended, and Mark welcomed me back onstage. The roaring applause that lasted for five minutes was unbelievable and seemed out of this world! Obviously, the audience had connected in a very deep way with the elephants portrayed in the film. This effect prevailed wherever the film was screened. Maybe, this is due in part to the fact that we have in our human brains a genetic chip that remembers when we, as a young species, interacted with animals such as these pachyderms. The bond that exists between man and elephant is an extremely primal one.

The outpouring of emotion was staggering. As a result, we were quite successful with our fundraising efforts that night. In the end, the minutia that I had focused on during the planning stages of the screening did not really matter. The incredible standard of the film and its emotional narrative left an indelible impression on people's hearts. These screenings and similar reactions of the audience told me that, no matter what, I was on the right track. The film was really impacting those who saw it. This was firm confirmation for me to continue to forge ahead.

The same evening, I had an urgent e-mail from a journalist in Kerala. Dhinesh informed me that he had spoken to the former speaker of Kerala's General Assembly (KGA), and he was open to discussing a possible screening of the film on the KGA grounds. I was still coming off the high that the LA screening had created and, thus, was a bit distracted. However, I e-mailed Dhinesh the pertinent information about the movie, the movie trailer, and a synopsis of *Gods in Shackles,* and then moved on to other pressing matters before travelling to India the next day. I was nervous, of course. Once again, as I had done so many times in the past, I swallowed my fear and tried to surrender to the higher power who I knew was really running the show.

Chapter 17

Miracles in India

My plan was to fly to India and stay there for a month to conduct screenings in various parts of the country, beginning with Kerala. However, I was so engrossed in the production of the film that I had neglected to finalize a working relationship with a public relations company, or rent auditoriums to screen my film, or even book my own accommodation for that matter! I had also tried to schedule film screenings in local theatres, but most of my e-mails and phone calls to the managers went unanswered. The previous month, I'd also personally invited some of the newly elected officials, announcing my film release. But it felt like we were speaking an alien language as I struggled to understand their English, and some spoke only the native tongue. In short, nothing seemed to be working!

In just a few hours, I'd be leaving for Kerala. But I began to feel overwhelmed and confused and struggled to breathe. Maybe I shouldn't go? I was gripped by fear, terrified of the consequences of screening a highly provocative documentary exactly where it had been filmed. The controversy surrounding *Gods in Shackles* was such that I would require a security detail not only for my own safety, but also for the attendees, just in case things got out of control.

The film was a grave threat to elephant owners, brokers, corrupt veterinarians, and even the temple authorities as the truth would be exposed. And these people would stop at nothing to muffle me and my message, fearing that it could potentially lead to a collapse of their dynasty, and even the entire 'elephant entertainment industry'. I became a victim of cyberbullying, with trolls hounding me on social media, trying to tarnish my work by calling me an enemy of the culture and a foreigner who had abandoned Indian traditions, even as these detractors made false claims that captive elephants were being well cared for by their handlers. I had also received dire warnings to stay out of their business, or else . . .

The flagrant lack of concern and basic dignity for animals or humans is the root cause of cruelty that causes unnecessary suffering to both humans and elephants. The lobby groups resort to spewing hatred and violence against those who speak out so they can retain their control and financial interests. And now, I could experience the terror and threat that the elephants feel.

Dark circles began to swirl in front of my eyes. I was suddenly reminded of Lakshmi's anguish as the mahouts popped out her eye, but hard to imagine the darkness that must have consumed her in that moment. We were in this together now. But the big difference was that I was being protected and cared for in ways that she had never experienced. With so much pain and guilt in my heart, I felt it would be best to take a step back and postpone my trip. Just minutes before leaving for the airport, I kept phoning the airline like a madwoman, and hung up each time the operator answered.

Frankly, the thought of security had not even crossed my mind. It was only when Paul Lewis, my media advisor, had brought it up that I even considered its necessity. I had been living and breathing this film for three years, and it had consumed me. Paul reminded me that I'd been so focused on exposing the plight of the elephants that I had neglected my own safety. Perhaps my childhood conditioning was a key factor in perpetuating this habitual disregard for my own self. But then I was not alone.

Women and elephants are also trained at a very young age to suppress their innate nature and serve men, without challenging their authoritative demands. Even today, girls are quickly silenced when they confront the authority and are forced to do what's told. If not, hell would break loose, and the woman or girl would get yelled at or even beaten up. Some of the misguided cultural myths are responsible for the afflictions that perpetuate violence in a male-dominant society.

My parents had drilled me to be humble and subservient and to put other people's needs before my own. This is probably why a 'selfish' thought such as arranging a bodyguard hadn't even crossed my mind. Perhaps I didn't want to draw attention to myself. Or maybe deep down, I believed myself to be unworthy, given that I had never really felt valued or appreciated by my own parents. But regardless, in that moment I realized that I had strayed too far over to the 'selfless' end of the spectrum. I wasn't applying for sainthood, after all! It had been one thing to be nervous about talking to the censorship board in India, but this was something else entirely. My own physical safety was at stake, and I would need all the spiritual tools in my toolbox to overcome my fears.

I instantly dropped everything I was doing and sank into my couch. With my eyes closed, I began to inhale and exhale deeply. I felt like an isolated and scared child. I began to hear voices in my head. One pushed me to cancel the trip and the other one said, 'Well, what have you got to lose? The worst-case scenario is that they will hurt or kill you, but this will only draw more attention to the plight of elephants.'

Eventually, after I let the whole thing run its course, my paranoia subsided. While the security threat was real, I realized that my fears had more to do with the rejection and disapproval that I had experienced most of my life. My feelings of inadequacy were manifesting as fear. But when I had ascertained the root of my fear, I began to journal about my delusions and the doubts that clouded my mind. And as I began to journal, my anxieties were gradually released. After a few moments of this practice, my trust in fate's unfolding was restored.

I landed in Trivandrum on 23 June 2016—the exact date of the third-year anniversary of my father's death. You might recall that in June 2013, I went into the jungles of Wayanad with Raj to rescue the bull elephant that had fallen into the trench. This is where my entire journey had begun. Exactly three years later, here I was, having come full circle. The timing could not have been more perfect. I was acutely aware of this synchronicity. It indicated to me, once again, that I was indeed on the right path, and this realization gave me profound solace.

The same evening, Sarita, my public relations officer, came over to my hotel to meet with me and strategize the screenings in Kerala. I was deeply touched by her commitment to the cause. This appreciation enhanced after I learnt that she was tapping into her personal resources. She involved her mother, who had worked for the government and was now harnessing her connections to help us promote the film.

It's important to mention here that a new government had been elected in Kerala just a month before the movie screening in June 2016. The initial plan was to release *Gods in Shackles* in December 2015. However, due to the seemingly unending challenges during the editing and post-production stages, the release date was pushed back to June 2016. The specific screening date had yet to be worked out.

I now had to hit the ground running! The next morning, I met Sarita and her mother at the government building. Here, I was introduced to key bureaucrats, but the ministers were unavailable. Watching the way Sarita and her mother navigated the system underscored the necessity of building trusting relationships on the ground in order to get things done. My fears about not having security seemed unfounded. The voices in my head had tried to thwart my flow. In reality, however, few people in Kerala even had a clue that a movie about temple elephants was about to be screened. I talked myself out of being paranoid and reminded myself to cool down. This space and time allowed me to relax and refocus on my mission.

Then 26 June arrived. Through Dhinesh, I had managed to schedule a time to meet the former speaker of the Kerala Legislative Assembly, Sri Radha Krishnan, who offered to introduce me to the newly elected speaker, Sri Ramakrishnan. I was intent on asking them if they would consider screening the film at Kerala's Legislative Assembly. If they agreed, the ramifications for the elephants could be awesome. The policymakers convene here on a daily basis, and watching the film could potentially lead to major policy changes.

We met both the speakers of Kerala's General Assembly at the appointed time. They had been made aware of my film and, no doubt, about my mission as well. I began by telling them that the movie was not flattering for Kerala as it exposed the cruelties meted out against elephants. I also explained that the film would be shown across India, and it would be prudent to unveil it in Kerala, as the issue pertained to their state. This would be an acknowledgement of the temple elephant abuse there.

I also made them aware that the world was appalled by the elephant abuse transpiring in Kerala, amplified by the prominent film festival circuits where *Gods in Shackles* had been featured. I then made the most significant ask, 'Would you, sir, the hon'ble speaker, consider inaugurating the film at Kerala's General Assembly?' He looked at me intently. His expression was tough to read, and it was hard to tell if the urgency of my message had been properly conveyed. Finally, he nodded his head and invited me to his office the next day.

The following morning, Sarita and I promptly arrived there at the specified time. With us were an official request and a copy of the film. We were informed that the speaker was in session. He, nevertheless, had arranged for his press secretary and personal secretary to screen the movie. As the press secretary watched *Gods in Shackles*, his facial expressions revealed shock and horror. Every half hour, he would get distracted by someone, and leave the office. It took him nearly three hours to finish the ninety-two-minute-long film. But in the end, he said, 'The truth has to be seen by everyone,' while also cautioning that 'Madam secretary has to decide.' And more nail-biting moments ensued. Even if we secured her approval, the speaker's decision would ultimately prevail. At this point, the secretary left the office to meet the speaker and to provide him her impressions of the film and recommendations.

I began to pace back and forth restlessly as the staff kept me supplied with hot tea. They must have refilled my cup at least three times! The fourth time they offered tea, I politely declined, thinking: 'I don't need any more tea, I just need an answer.' Given my acquired North American temperament, I began to feel a bit impatient. I suspect that a few sarcastic comments had slipped out of my mouth. Finally, at around 5:30 p.m., the personal secretary re-entered the office. The serious look on her face made me wonder what had happened. After a few moments, she broke the news. The speaker had agreed to show the movie on the assembly grounds on 29 June. I could not have asked for a happier ending to a tense day.

But the excitement lasted only a few minutes. How would it be possible to pull this off within such short notice? In two days, we had to print invitation cards and deliver them personally, as is customary in India. Fortunately, our graphics designer in North America hurriedly created an invitation. We printed the cards the following day and delivered them to the speaker by 3:00 p.m., giving him enough time to distribute the invitation to the members of the legislative assembly (MLAs). I then asked the personal secretary to take me to the banquet hall to survey the space and equipment we would use for the screening.

The room was totally unsuitable, so the speaker decided to screen the film in a different building of the legislative assembly grounds. It suddenly dawned on me that the invitation the speaker had handed out to the MLA's had the previous location printed on it. Realizing the situation, the speaker made an announcement in the General Assembly, in what was another unusual move, as it drew the media attention. Soon, show time arrived. The place was packed with at least three hundred people, many of whom were MLAs and otherwise very influential people. It also included a presidential award-winning film director, forest officials, members of animal welfare groups, cable networks and print media outlets, and most importantly, the speaker himself. The anticipation was building as I went around the room, greeting everyone.

As the speaker took to the stage, replete with a podium and microphone, he invited me to sit right next to him. His personal secretary introduced the speaker. And he spoke from his heart about the social issue that I had raised in the film before introducing me. Trying to maintain my composure, I welcomed everyone and introduced the film. In a matter of moments, the screen was lowered, and *Gods in Shackles* rolled.

Gods in Shackles made history in Kerala as it was the first film ever to be screened on the legislator's grounds. It also made news headlines, creating unprecedented awareness about the plight of Kerala's captive elephants. It continued to make the local and national news for several days thereafter.

Looking back on this poignant day of my life, I could have never dreamt that this would happen in Kerala. I had arrived in India with such trepidation, but now I had all the reasons to celebrate, for the film had afforded me a pulpit to be a voice for the suffering elephants due to the extensive press coverage. Quite the contrast to my childhood, when my voice and opinions were suppressed. It was difficult to reconcile this dichotomy, and it took me a while to integrate this new reality. Like the elephants featured in my film, I, too, was in the process of consciously becoming unshackled of my inadequacies.

The next day, Sarita and I visited the Kerala State Film Development Corporation (KSFDC). Here, we booked theatres for back-to-back public screenings in the cities of Thrissur, Calicut, Kochi, and Trivandrum—all in the state of Kerala. It was midday by the time we left. Given that the film had made a huge media splash and received unprecedented publicity, I was advised that I would need a security detail for these other screenings. To this end, we were scheduled to meet the speaker again to discuss police protection in Thrissur, the epicentre of the elephant-entertainment industry.

When we reached the speaker's office, he was in the General Assembly Hall, and it wasn't until 4:00 p.m. that we got to see him. I then explained the purpose of my visit, stating that I was travelling to Thrissur the next day and requested for police protection. He immediately hand wrote a letter in Malayalam, Kerala's native language, attached it to my formal request, and made a couple of phone calls. He then handed me an envelope and asked me to take it to the chief minister's (CM) office right away.

At the CM's office entrance, three policemen and two undercover officers in plain clothes afforded me access. Behind a large wooden table was a man with salt-and-pepper hair and glasses, wearing a white shirt and white sarong. I was standing before the most powerful man of the state—Sri Pinarayi Vijayan, the honourable chief minister of Kerala who had come into the office only a month prior. He oversaw the welfare of Kerala's 36-million-plus people. As I handed him the speaker's letter, my hands were trembling just a bit. He read it, then looked at me and smiled, with a gesture of approval and reassurance.

That same evening, I went to the police headquarters to co-ordinate details of my upcoming travels with them. Although the state police chief (SPC) was from the state of Orissa, he spoke Malayalam very well. He commanded people's respect, for he honoured their culture and traditions. During our informal chat, the SPC conceded that the use of elephants in temple ceremonies was proving to be a significant public safety and security threat for the society. He concurred that banning elephants in festivals would have to happen gradually. Meantime, a commando had booked my ticket to Thrissur, and the police superintendent had arranged two personal security officers (PSO) to escort me during the screenings across Kerala. I was advised that they would travel by train and meet us in Cochin.

As we drove from the airport through the congested roads, memories of filming with Tony flooded my mind. I recalled the hot and humid air, the palm trees, the canals, and the people. I also recalled the deeper insights I had been privy to when managing my impatience with Tony. In the next moment, we approached the famous traffic circle in Thrissur. Here, poignant memories of my precious Lakshmi flashed back before my eyes. I glanced at the exact spot where this elephant had paused in reverence to the dead cat as I'd watched the oblivious drivers driving over the cat's little corpse. Suddenly, tears began to roll down my face. My life had changed so much, and my perspectives and world views had shifted dramatically. It was all for the best but the feelings that were now arising within me were indeed intense.

It was too late for dinner when we arrived at our hotel, so we went straight into our rooms. But I was too excited to sleep, as I would be seeing my beloved Lakshmi the next day. Early next morning, the police escorted me to the temple. From being terrified of not having protection, and almost cancelling my trip for security reasons, here I was being protected every step of the way. I only wished they would provide, at least, the same kind of protection, if not more, to the voiceless elephants and somehow alleviate their silent suffering.

My heart skipped a beat, as I stepped into the temple at around 6:00 a.m. The familiar oil lamps and the sounds of temple bells filled the air, but there was a sense of emptiness amid all the rituals. I turned to the spot where Lakshmi was usually tethered. It was vacant! I consoled myself, maybe she was running a bit late. But usually, she always arrived on time for a special morning worship service called *seeveli*. It was her job to carry the priest and the statue of the deity on her back and make three circles around the altar as the devotees followed her. But where was she today?

Men in white sarongs and women in saris were praying with their eyes closed, some prostrating before the altar, and others confused by the heavy police presence. As soon as I entered the altar area, the highest priest lit a tall lamp with almost thirty flames. The bells chimed as he gently moved the lamp in circles around a deity adorned in flowers and jewellery, greeting the day. The priest then handed me some flowers, sandalwood paste, and sweet rice pudding wrapped in a small piece of banana leaf. A trickle turned into a stream, with tears flowing uncontrollably. I circled the two altars inside the temple, praying for only one thing—to see my beloved Lakshmi. I was also keenly listening to the haunting sounds of shackles that would announce her arrival. But not a trace!

Almost forty-five minutes later, my PSO said that we had to leave. For one last time, I turned and looked at the space where Lakshmi was usually tethered. It was empty, just like my heart, and it suddenly hit home that I would not be seeing my sweet Lakshmi that day. Like a madwoman, I desperately began asking everyone where she was. Finally, one of the temple authorities approached me and broke the devastating news—she was unwell and would not be performing *seeveli* that day. Apparently, she was undergoing a special treatment for a month. It felt like a huge boulder had been dropped on my head, and I stood there numb and speechless.

During every visit to the temple, Lakshmi's presence had filled up my heart with love, joy, and happiness. She compensated for my feelings of inadequacy and loneliness in the solidarity I felt with her. Lakshmi also suffered from loneliness, which was imposed on her by her physical confinement. To say that I identified with her suffering would be an understatement. My PSO tried to console me. But my heart was shattered into a million pieces. I struggle to find words to describe the heavy dark feelings I experienced that day. It was hard to remain composed.

With my head hanging down, I walked towards my car. But I wasn't ready to give up just yet. I asked the policemen if they could accompany me to Lakshmi's place. However, they said that she was chained in Sundar Menon's backyard, a private property that can't be accessed. The yard of the mansion where Lakshmi was tethered has two ferocious German shepherds whom I had encountered several times during my past visits. And the place is protected like a fortress. The police did make a few cursory phone calls but to no avail. What would I do with all the fruits I had bought for her—the pineapples, watermelon, and her favourite bananas? All I wanted to do was feed and hug Lakshmi. But it was not meant to be.

I suddenly realized that it was 8:00 a.m., and I had to be at the Kairali Theatre by 9:00 a.m. for a test run before the public screening at 10:00 a.m. I quickly swallowed my emotions, and putting on a brave face, called the theatre manager to confirm that we would arrive on time. We then drove back to the hotel where I dressed for the part, and off we drove to the theatre. The security was tight as we were in Thrissur, the ground zero of elephant torture.

Near the theatre's parking lot, heavy barricades had been erected to keep the public at bay, and at least fifty police officers had been dispatched for controlling the crowd. Snaking through the crowd, a man with silver hair and beard, wearing a white shirt and sarong greeted me with a gesture of namaste. He was none other than the activist Venkitachalam, featured in my film, and I was delighted to see him. A huge crowd tried to follow us as we walked through the theatre entrance, but the police blocked them. As soon as I entered the projection room on the second level, I peeked down into the auditorium for a quick glimpse, only to discover a few men conducting bomb tests. I tried not to get distracted, although it was frightening to even think that people could be so fanatical as to blast the entire theatre—so controversial was the release of *Gods in Shackles*.

After a quick check in the projection room, I entered the auditorium just before showtime. Dr Cheeran and Dr Giridas, two veterinarians with opposing views featured in the film, were part of the audience. Dr Giridas is the same vet who had said that one of my questions was 'stupid'. I noticed a smirk in Dr Giridas' smile and later learnt that he was there to witness the drama that could potentially unfold between me and the elephant owners. But there were neither bombs nor conflicts throughout and after the screening. Thanks to the Kerala police!

Soon after the movie ended, Dr Giridas scurried out of the auditorium, realizing that the views he had expressed in the film did not reflect well on him. Unable to handle the truth, he escaped the media. Meanwhile, people came rushing up to me with tears in their eyes. One elderly man, in particular, was so deeply touched that he placed his hands on my head, whispering, as an elder would, 'May Lord Ganesh be freed from the shackles of suffering.' It was indeed a tender moment.

The minute I stepped outside the movie hall, spotlights, cameras, and the media were throwing a barrage of questions. Suddenly, out of nowhere, a man came dashing violently through the crowd. He grabbed my right hand, and slapping his business card angrily into it, said, 'I'm promoting these gadgets!' The police officers swung into action and escorted him out of the building. He turned out to be an elephant broker, a middleman, who leased out elephants for huge profits. Venkitachalam, who was also present at the media scrum, explained that this broker was referring to a device invented to shock the elephants and inflict severe pain by simply pressing a remote button. It was designed to control the elephants using brute force to 'obey the commands' of the handlers.

Soon after the media interviews had ended, four men approached me and introduced themselves. They were the Kerala High Court lawyers who had travelled all the way from Cochin city of the Ernakulum district. They were working together to end elephant abuse and said they wanted to screen it at the Kerala High Court for judges and lawmakers. Apparently, a woman named Meera had asked them to screen the film to assess its veracity. Meera had contacted me via Facebook a few months back. However, in my busyness, I had read her e-mail but hadn't responded. She persevered, and it was heartening to see that these lawyers earnestly wanted to end elephant abuse.

Later that same afternoon, we travelled from Thrissur district to Calicut district for the second screening on 3 July 2016. More than four hundred people packed the auditorium, including youths belonging to the local police chief's club. Again, the sniffles and sobs that the film evoked in people could not be ignored. Somehow, in some inexplicable way, the audience members had bonded deeply with the suffering elephants depicted in the film. On top of the deluge by the audience was the standard media scrum. After that, I was escorted to one of the most popular TV stations, Manorama. It had regularly been covering the atrocities against the elephants. Through it all I learnt the value of human relations, building trust, celebrating everyone's efforts, and empowering people. It was incumbent on them to help make a difference, too.

We flew back to Trivandrum the same day to make arrangements for a public showing there on 6 July. Early the next morning, one of my police escorts drove me to the home of Kerala's most revered poet laureate, Sugathakumari. I hadn't seen her since I'd interviewed her in June 2015. This would be a reunion of sorts. At the door, her sister informed us that the teacher was recovering from a hip surgery. As we entered her room, she was lying in her bed and sounded meek. Her first question was whether I would be producing a Malayalam version of *Gods in Shackles* and have her narrate the story. This legendary activist has been relentlessly fighting for animals, women, and Mother Nature for decades. And now, she was about to create a lasting legacy for elephants by lending her powerful voice to the elephants. After extending the invitation card personally, I returned to the theatre to ensure that everything would flow smoothly at the screening.

Finally, the big day arrived. One by one, the dignitaries showed up at the Kalabhavan cinema complex. As I welcomed everyone, my eyes kept searching for Sugathakumari, whose presence meant the world to me. Suddenly, a white car arrived, and my PSO realized who it was. All heads turned. I hurriedly dragged a wheelchair right next to the door and wheeled her to the VIP balcony where the dignitaries and the media had convened. The minute she arrived, she was instantly swarmed by reporters and cameras, and she articulated clearly in their native language the plight of Kerala's temple elephants as exposed in the film.

It was a packed house with people sitting on the steps. The ground level was crammed with the student police cadets, and the balcony was brimming with dignitaries and activists from neighbouring districts. As the movie rolled, the nagging question that kept haunting me was whether it would propel conscious actions over the long term. But I had to silence my inner voice and remind myself that I could only control my actions and had to trust in fate's unfolding.

Another major screening in Kochi also drew the district collector, politicians, and activists. Here, I finally met Meera, who had arranged for the High Court lawyers to screen the film in Thrissur. Other community screenings in universities, colleges, and press clubs also took place on a smaller scale. They were all packed with curious students, the future generation of our planet. Some defended their traditions, but most were willing to do something to end the exploitation of elephants. I left Kerala having cultivated relationships with some of the key influencers of the state and making significant inroads, which caused me to feel a bit hopeful on the elephants' behalf.

I was now ready for some rest and respite. I travelled to Mumbai, where my heartstrings were called to heal my childhood wounds with my widowed mother.

Chapter 18

Shifting Family Dynamics Empower Me

As I was waiting in the airport lounge to board my flight to Mumbai, a strange sense of discomfort and angst set in. Memories of my short visit to see my ailing father in November 2011 returned to haunt me. What transpired during that time is a subject matter for another book. But suffice to say that I was feeling a bit uneasy to visit my mother after a three-year absence, not to mention that I was supposed to stay with her for ten days.

In less than two hours, the aircraft started to descend. I began to hyperventilate and tried to breathe deeply, trying to calm myself down. As I exited the airport, my brother was waiting to receive me, and, in a short time, we arrived at our mother's home. It was around midday, and my mum served us a delicious vegetarian south Indian lunch. My brother hung around for a bit, reminiscing his last few days with my father before returning to his offshore duties in the Middle East. Then, the moment that I dreaded the most arrived. My brother was leaving! Now, I had to spend the rest of the day with my mother. What if she made me feel guilty and ashamed for not attending my father's funeral?

For a few moments, there was an awkward silence. Then my mother began to pour out her grief, sharing my father's final moments, and how she noticed his eyes scanning around the room for me. She vividly depicted everything that transpired until his cremation, without projecting an ounce of guilt onto me. Her demeanour had changed since I'd last spoken to her from Toronto not too long ago. And she was interacting with me, as though we were long-lost friends. The tension immediately dissipated. My frozen stance began to thaw. Things were going to be okay.

Taking a deep breath, I smiled at my mother, with tears welling up. Then, I shared with her the news of my sojourns for movie screenings and all the wonderful people who had come along the way, and said, 'If Dad had been alive, he'd be proud of me.' To which, she innocently responded, 'He wants you to spend a few days with me,' and it brought a smile to my face.

For the next few days, our mornings were serenaded by an elephant-loving spiritual leader, Swami Bhoomananda Tirtha. His TV show called *Spiritual Splendour* resonated deeply as he used nature metaphors to explain the ancient Hindu scriptures of the Holy Bhagavad Gita. I had previously heard through various sources that he had been fighting for the welfare of Kerala's temple elephants. My mother admired him so much that she frequently wrote to him. In fact, she'd even tucked away his responses in a locker. When I told her that I wanted to meet him to discuss about the captive elephants, she promptly shared his contact information that she had gathered from the TV show.

Over the next few days, my mother taught me many things—some, of this world, and others, out of this world. For instance, she showed me how to make protein powder using high-fibre lentils to cook up a scrumptious and healthy breakfast. Occasionally, she allowed me to prepare her favourite dishes, although for the most part, she was in charge of the kitchen. We had so much to catch up on. Aside from mundane gossips about her sisters, neighbours, and my brother, my mum also read and delved into the scriptures of the Holy Bhagavad Gita. In this, my mother imparted her wealth of knowledge on the workings of the mind and lessons on letting go. I began to admire her simple life and her strong faith in God. My respect and admiration for her became unbounded. In turn, she showered me with beautiful words, blessings, saris, and most importantly, unconditional love.

During all of this, a light bulb suddenly went off in my psyche and something shifted profoundly inside me. I realized that I had surrendered, allowing the smooth flow of energies between my mother and me. I was practising living in the moment and being present during all of my exchanges with my sweet mother. I had definitely changed, and so had my perceptions. My preconceived judgements had melted away, allowing me to see things as they were, rather than interpreting the way my mind perceived them to be. I was becoming acutely aware that the stories cooked up by my mind were mostly false. After having discarded my tainted perceptual lenses, I began to see my mother through pure and innocent eyes. I listened to her intently, valuing her opinions and perspectives. All of these factors now created peace and harmony between us.

Before I knew it, ten amazing days with my mother were coming to an end. It was hard to leave her behind. But somehow, the world now seemed a friendlier place. When my relationship with my mother was healed, it became easier to heal other human relationships. When we heal our intimate relationships, we can create peaceful coexistence on the planet.

I left Mumbai, cherishing my newly found love and admiration for my mother. A heavy burden had been lifted off my shoulders through our open conversations that allowed us to heal. Surely, elephants had played a significant role in reuniting me with my mother, for if not for my screenings, I probably would not have visited my homeland, nor returned to Mumbai. In fact, since my journey into the making of my film began, these gentle animals had been aligning the perfect people and situations at the perfect time. I sensed that elephants were also mending my broken relationships, beginning with my brother's visit to Kerala, where I was hospitalized after my accident.

It's important to mention that these insights or healing did not happen overnight. It had taken decades of living independently in Canada, which gave me time and space to embark on a personal inward journey. I had to do it on my own, distancing myself from everyone. Although many westerners travel to ancient India for spiritual healing, I found my soul in Canada, after spending a great deal of uninterrupted time in solitude, communing with the natural world, and connecting with the creatures of the earth. Growth cannot be forced or hastened; it has to be nurtured. It may take months, years, or even decades, but a little bit of patience, willingness, a strong desire, and some discipline can do the trick. I confess that I am still a work in progress.

My next destination was New Delhi, India's capital city, where a major screening of *Gods in Shackles* film was planned at the Connaught Place, in the heart of Delhi's downtown.

As I approached the famous PVR Cinemas on the eve of the screening, a young lad came running towards me, on the verge of tears. 'Ma'am I am your Facebook friend and I'm here to see *Gods in Shackles*,' he said. 'But I didn't print out my ticket and they're not allowing me in.' Samarrth Khanna, a thirteen-year-old youth, had travelled by bus all the way from Faridabad with his mother. They had undertaken this two-hour journey to the heart of New Delhi just to watch the film and meet me. He couldn't understand why the ticket dispensers would be so mean to him. Deeply touched by his passion for the elephants, I held his hand and escorted him and his mother into the theatre. As I walked with my arm around his shoulder, he glanced at those who had denied him access, with a smirk on his face. No doubt, he was feeling vindicated and happy to show off his new connection.

As the film unspooled, there were shocked expressions on people's faces. This never ceased to amaze me. The national and international media coverage created a huge demand for screenings across India. Meanwhile, Dr Chinny Krishna, the former vice-chairman of the Animal Welfare Board of India (AWBI), had contacted me for a screening in Chennai.

This cultural capital of the state of Tamil Nadu is filled with temples where elephants are forced to stand on concrete floors. Here they must bless the temple devotees by touching their head in exchange for cash donations. To put it bluntly, elephants, considered to be the embodiment of Lord Ganesh, are made to beg inside the temples. Aside from the stark irony, this is a highly controversial issue. Scientists have been arguing for decades that hard concrete floors are simply unsuitable for the elephants' soft cushioned feet that are designed for lush jungles and marshes. But still, the cultural and religious sentiments often take precedence over the welfare of elephants.

Although the neighbouring Tamil Nadu state does not have nearly as many elephants as Kerala, the one hundred and fifty or so in the state suffer a similar fate. One big difference between Kerala and Tamil Nadu is that in Kerala, elephants are not forced to bless people by touching their heads. Secondly, most elephants in Tamil Nadu are females. In the wild, they often live in tight-knit herds, constantly socializing and communicating. But they are ripped apart from their families at a very young age, tortured to submission and then forced to stand on hard concrete floors only to beg in the temples. Not only does such violence causes physical wounds, but also psychological traumas. What a pathetic plight of a highly intelligent and regal species!

Soon, the movie rolled. After it ended, true to form, there was hardly a dry eye in the house. Although most people were profoundly touched, there were some deniers. They clung onto their cultural biases and beliefs despite the raw video evidence in the documentary.

The tension was palpable, particularly as a man tried to distort some of my responses during the Q&A. He told the packed auditorium that I was tearing apart the cultural fabric of India and brainwashing people to boycott its temples. What I had actually said was that people should boycott the festivals that exploited elephants and speak out against the way elephants are brutalized in the sacred temples.

The entire event lasted for almost four hours. When I was leaving, utterly exasperated by the unnecessary friction at the end, a young woman named Janani introduced herself. She was on the board of several animal rights and wildlife organizations. She also managed an animal welfare organization in Kodaikanal, rescuing stray cats, dogs, cows, camels, and even elephants. She undoubtedly drew criticism, as standing up for animals and fighting against a cultural norm is not easy in India. In this nation, many people still believe that the human race is superior and that animals have reincarnated in that form because they have been cursed for their past life sins. But Janani is one of a kind. In the ensuing months, she would become a key ally and host numerous screenings to spread awareness across Tamil Nadu.

The following morning, as Chinny, his wife, and I sat around the table for breakfast, they cautioned me to be vigilant of people like the man who had tried to stir controversy at the screening. Sure enough, in less than a week, he had published a scandalous review about *Gods in Shackles*. I ignored it for a while, but eventually I wrote a rebuttal, defending my stances against elephant abuse and fending off critics who accused me of ripping apart India's cultural fabric.

I was headed next to Bangalore for another screening, hosted by my kindred spirit, Suparna Ganguly. Without her support, none of the Indian screenings would have taken place, to say nothing of her prominent role in my film. For the next few weeks, there were, at least, three news stories every day about *Gods in Shackles*, drawing national attention to the plight of Asian elephants.

By the time I returned to Toronto, requests for screenings began to pour in from North America. The monumental June screening in Los Angeles also had a cascading effect. Additional screenings were held in Baltimore and San Diego and both of them were extremely well received. Susan Ciaverelli and Danette Shue, two dynamic women and animal lovers had joined hands for a common cause, and they put on a fantastic show at Baltimore's historic theatre, *The Senator*. Inspired by the Los Angeles screening, Devvie Deany, another passionate animal rights advocate, had organized a showing at the most rustic auditorium in San Diego, drawing hundreds of people. The relentless efforts of activists around the world made a huge impact, with the audience asking to volunteer to become part of the solution. All in all, we ended up raising substantial funds, every penny of which helped to co-ordinate more screenings in India.

Clearly, money was just a means to an end—in this case, to create awareness and education. But more significant were the heart-to-heart conversations with hundreds of people in theatres and arenas. My new friends—the elephant allies whom I had met through Facebook—all joined hands in what was rapidly becoming a global movement. The screenings brought strangers together who then became friends, united for a common cause. In this, the results of their efforts were compounded. Each screening had given me a deeper understanding of how the universal forces were working.

When I reflect back on the entire journey, I can't imagine what would have happened had I allowed my trepidation and anxiety to prevent me from embarking on my flight from Toronto, and doing what I knew I had to do. A higher power always redeemed me from my doubts and fears by giving me the strength and courage to move through them.

Chapter 19

Marvels Abound

I returned to Toronto with an overflowing heart to find another miracle awaiting me there. *Gods in Shackles* had been selected by the International Film Festival of India (IFFI). This festival would be held in Goa in western India on 26 November 2016. IFFI is India's only official film festival and considered to be the most prestigious. I had planned a return trip to India for 23 November, and the timing couldn't have been more perfect!

Unlike most films that were shown only once at the IFFI, *Gods in Shackles* was screened in the morning and afternoon of the 26th to a packed audience. It included Bollywood actors, Goa's forest department officials, and other delegates. Many young people in attendance were from Kerala. Inspired by our meeting in Goa, at least a handful of them went on to broadcast the film at their colleges and universities. The message began to spread like wildfire as the voices to end elephant abuse was getting louder.

This screening left me elated and highly energized for my next journey. I travelled to Colombo, the capital city of Sri Lanka, directly from Goa on 28 November. Here, I met Sujeewa and Sue, my gracious hosts. Varma, the elephant researcher, whose studies on captive elephants had inspired me to produce *Gods in Shackles*, had connected me with them in early 2016. They were thrilled to bring the documentary to Sri Lanka where Asian elephants face the same cultural exploitation that they do in India.

The first national survey of elephants in Sri Lanka revealed that there were approximately six thousand wild elephants, with a fraction of that number in captivity. In Sri Lanka, just like in India, these elephants are paraded in religious festivals and exploited in the tourism industry, forced to give elephant rides on their delicate spine. In Sri Lankan temples, Buddhist monks had been blamed for this wrongdoing that gained notoriety among the animal rights activists. And as in India, the issue had turned highly controversial.

One particular elephant named Ganga had become the poster child for elephant abuse in Sri Lanka. Monks had come under fire for the barbaric treatment of this female elephant. She had been deprived of the most basic necessities of life. Sujeewa and Sue were on the front line, leading several effective campaigns and court battles to end elephant abuse. So, the screening in Columbo was a monumental step for Sri Lanka. It was a packed screening with the usual Q&A and other formalities.

The next day, we set out to a protected wildlife zone called Uva, not too far from Colombo. The sun's intense rays penetrated my skin and the humid air made it difficult to breathe. We drove past an open area where a wild bull elephant was happily grazing beneath the scorching sun. Not too far away was a dam where he could take a dip to cool off. He was free to do whatever he chose. But most captive elephants are doomed to misery until they die, with only a lucky few rescued and rehabilitated by earnest people.

In about twenty minutes, the atmosphere changed suddenly as our jeep had veered into the magical Uva jungles. Shortly thereafter, we spotted two herds of majestic wild elephants. The one closest to our jeep had five adult females and a baby, who was just a few weeks old. It was suckling its mother's milk, as the others in the herd shielded the little one. It was awe-inspiring to be with these animals, watching them touch each other and lock their trunks playfully together.

Further down the road were four jeeps packed with tourists, witnessing another herd of elephants peacefully grazing. One of the drivers audaciously pulled his vehicle very close to the second herd of elephants to impress the tourists, and potentially earn some extra cash. Neither the driver nor the tourists seemed to care that they had encroached into their space and agitated the matriarch. She aggressively dashed towards the vehicle, and a standoff ensued. But the driver refused to retreat and so did the matriarch. After a few minutes though, she backed up, realizing that her herd had returned into the jungles, along with the baby.

Our next destination was the southern province. As we approached a place called New Tangalle, Sujeewa asked the driver to pull over near the wall of a large temple, and he hopped out. I had no idea what he was doing, but I blindly followed him. He then jumped on the wall and signalled me to join him. Below was a bull elephant tethered in cruelly short chains, swaying in distress, no roof to protect him from the merciless heat. Farther down in the temple's yard, a female elephant was tethered inside a water tank. She was also unable to move. But, at least, she could cool off. They looked depressed and neglected.

Elephants have the largest brain of all animals on land or in water. They are always active, utilizing their intelligence to create tools and engage in activities with other members of their herd. But here, they were bored out of their minds, with nothing else to do. It was appalling to realize that the same Buddhist temples that preached kindness, peace, and compassion were depriving these cultural icons of the most basic necessities of life! It was no different from what I had witnessed in Kerala.

The next day was a bit more inspiring. We visited a semi-wild transition home. Here, rescued baby elephants were being nurtured near a forest area to be gradually integrated into the wild. We arrived by 6:00 a.m. to witness an amazing spectacle. At 6:30 a.m. sharp, fifty to sixty baby elephants had lined up at the entrance for their morning milk. Five men were filling several buckets with the milk formula, while two others guarded the entrance. Once the bottles were arranged, the men at the entrance opened the gates. Five or six batches, each comprising of around ten teenage elephants came running for their milk all at once, acting very much like young people when they are hungry. They were then herded back to a semi-wild open area, so they could graze together and socialize.

There were also a few elephants in sheds, where some of them were rescued by Sue and Sujeewa. Apparently, they'd been tipped off by an undercover agent as the baby elephants were being kidnapped from the forest. They immediately alerted the police, assisting them in arresting the perpetrators, and ensuring that the babies were rescued. Thankfully, these elephants didn't end up in temples. But it was still heartbreaking to realize that they'd been ripped apart from their families by greedy men to sell them for loads of money.

The same compound housed a young elephant with a prosthetic leg, and a two-week-old baby elephant suffering from serious diarrhoea, who was hooked up to an IV line. His handler was gingerly holding the drip and allowing him to walk around. Amid all the heartbreaking stories, human compassion shone through in this semi-wild transition centre where poor people dedicated their lives to protecting these elephants. It was heart-warming to witness the special bond they'd cultivated with the baby elephants, caring for them like their own children.

That evening, Sue and Sujeewa had organized another major screening of *Gods in Shackles* in the central province of Kandy, the epicentre for elephant torture. Needless to say, the audience was shocked and appalled to witness the harsh realities that elephants suffered behind the glamorous decorations. The next day, a local hero had arranged a screening in a Buddhist school in the southern province.

Subsequently, students were given natural science photo books that were filled with images of native and endemic plants, birds, and animals, including the elephants. And then the Buddhist monks invited us for some high tea and Sri Lankan snacks. It was indeed inspiring to spend the entire morning and afternoon with the future generation of our planet.

But despite all these awareness programs, the religious institutions across Asia continue to parade elephants. No doubt, such abhorrent treatment of one of the most intelligent animals is intolerable, and we all want to see rapid changes. However, to put things into proper perspective, these practices have existed for several decades, and it could take the same amount of time, if not more, to change them.

One of the most distressing places I visited was the Pinnawala Orphanage, where baby elephants, who had been rescued decades back, now lived. I had heard some heart-wrenching stories of babies being sold to western zoos, as though they were commodities. The pathetic plight of an elephant named Lucy is a prime example of the atrocities that these animals are subjected to in this orphanage. When she was just a year old and still breastfeeding, she was ripped apart from her mother from the warm, tropical jungles of Sri Lanka and sold to the Edmonton Valley Zoo in western Canada in 1977. Now, 45-years-old, Lucy is forced to endure bone-chilling temperatures and bitter winds, and she suffers from several physical and emotional ailments, including arthritis and depression. Meanwhile, as activists continue the fight to have her released into a sanctuary, lonely Lucy suffers silently.

Such brutalities are cleverly masked by this Sri Lankan orphanage, where thousands of tourists from around the world flock to be amused by these young animals. When we arrived at the Pinnawala camp, the adult and sub-adult elephants were being taken to a nearby lake for bathing. They were then shackled to the metal hooks, deprived of the freedom to bathe comfortably. The handlers constantly threatened them with long poles that had pointed metal spears at the tip and used tremendous force to keep these gigantic animals under control. Most tourists were so mesmerized by the elephants that they barely noticed the torture that these elephants routinely endure. This is not the kind of orphanage that any country should emulate.

However, around the same time, I had learnt that the Kerala forest department was planning to emulate the Pinnawala model at the Kottoor Elephant Rehabilitation Centre, where Rana, Arjun, Poorna and the other elephants have a better life. I immediately contacted the chief wildlife warden of Kerala and sent him images of the ruthless treatment of elephants, hoping to shed some light on the ground realities in Sri Lanka, and how this could impact Kerala's reputation.

The next day, we drove to one of the most popular tourist destinations in the state. This was the province of Sigiriya where we would spend a few days. Deep inside the province, Sue and Sujeewa had set up the Centre for Eco-cultural Studies in a place called Diyakapilla. It felt surreal driving along dusty and bumpy roads enveloped by canopies of trees. There were also patches of human settlements along the way, featuring rustic houses with mud-tiled roofs supported by pillars made of tree trunks.

Just as dusk settled in, we reached our accommodation—a brand-new three-story building in the heart of the jungle. It was so newly built that we could still smell its cement setting. Unadorned glass windows covered half of the walls without any drapes. Sue and Sujeewa had moved in just weeks prior with only the bare necessities. After three intense days of screenings and travels, we could now relax for a couple of days.

Early the next morning, I was awakened by curious monkeys peering through the large glass windows, making me feel as if I were in a safari lodge. I opened the window and said, 'Hello.' And as though they couldn't believe their eyes, they looked at each other totally confused. They sat on the branch for a while, as I silently observed them grooming and communicating with each other in their language. Occasionally, they glanced at me just to make sure that I wasn't up to something, but then they carried on with their business as usual.

After a shower and my yoga routine, I went for a short walk. Outside, I was greeted by a cow and a bull, eating hay behind the fence. As I observed them, Sue walked up carrying brushes and combs. A good hour of grooming was followed by releasing the animals in an open, grassy area. Suddenly, the bull became playful and began to run, the cow pursuing him. Sue chased after them for ten minutes, yelling at the top of her lungs, trying to bring them under control. We were terrified that they might run away and hurt someone with their horns. As though the bull finally realized his folly, he surrendered, and so did the cow. It could have been disastrous.

I walked back through the slushy farmland, completely shaken, and sat on the easy chair on the verandah. Suddenly, chaotic scenes of elephants running amok in Kerala returned to haunt me. Unable to withstand the relentless sounds of music, revellers, fireworks, the intense heat on hot tar roads, and worst of all deprivation, even the most docile elephants are pushed beyond their limits. Every year, several elephants and people get killed during these festivals. But still, the government is unable to penalize them as temples are self-governing and autonomous bodies running their own agenda. It can stoke cultural and religious sentiments and even trigger riots and violence.

I was yanked back to reality by the same family of monkeys I had interacted with earlier that morning. They had descended on the patio from the trees, sitting a few feet away, and scrutinizing me. Just as we were getting to know each other, Sujeewa announced that he had arranged a tour guide to take me to the world-famous Sigiriya Lion Rock. This historic site features an ancient column of rock that had been used as the grounds of a palace by the fifth-century king. It later functioned as a Buddhist monastery until the fourteenth century. Today, it is a UNESCO World Heritage site.

With a few bottles of water and some snacks, my guide and I were soon off in our *tuk-tuk* (a three-wheeler). The place was packed with tourists. Snake charmers displayed live pythons wrapped around their necks. Tiny boutiques sold trinkets and memorabilia. The tour guide and I pushed our way through the commotion to begin our ascent of the huge rock. We reached the apex in twenty-five minutes. Once there, I glanced around. White puffballs were floating across the sky, and in the far horizon behind misty mountain ranges, the sky kissed the earth. From the highest point of the massive rock, I was relishing the tiny green patches of open space, wondering how many elephants might be wandering there.

After my rock-climbing adventure, we packed it in, returning to Colombo for yet another screening the next day, before flying back to India. The three of us had met as strangers but by the end of my stay in Sri Lanka, we had become kindred spirits. Our friendship was woven together by many common threads—culture, conservation, elephants, and food. Through the entire journey, I also observed how Sujeewa and Sue were getting so much accomplished by remaining relatively obscure. This was a trait necessary to manoeuvre through the bureaucratic processes in countries like India and Sri Lanka. Sue and Sujeewa lived a simple life, yet they seemed content and happy, serving a grander purpose.

On my final day, Sujeewa had arranged an interview for me at Sri Lanka's national television network. They said that they were interested in translating *Gods in Shackles* in their local dialect and airing it. It was a huge boost for our elephant movement that had now travelled to the most unexpected countries! The same afternoon, I met the chairman of the television station. Later, I was invited to provide a workshop for journalists and producers on how to gather powerful footage and create narratives that would resonate with the masses. This was another great opportunity to build bridges and foster collaborations to end elephant torture.

It was around midnight when Sue and Sujeewa dropped me off at the airport in Sri Lanka, and just past midday when I landed in Chennai. Janani had arrived to pick me up; she's the young woman I had met at the screening organized by Chinny and the Blue Cross Society. Janani had been so inspired that she had organized several screenings across the state of Tamil Nadu in universities, rotary clubs, and similar venues.

That night, Chennai experienced one of the heaviest torrential rains on record with a terrifying forecast for the following day. Tornadoes and thunderstorms with hurricane-force winds were in the cards. We drove through the heavy downpour, manoeuvring through the roads littered with massive trees that had been toppled down the previous night. She had arranged her driver to drop us off beyond the danger zone so we could then drive up to the Kodaikanal hills for our next screening. By noon, we had reached our destination, narrowly missing the eye of the storm by a couple of hours. A state of emergency had been declared in Chennai. There were significant power outages and destruction across the capital city of Tamil Nadu. However, we had escaped the worst.

We were now driving up the lush Palani hills to Kodaikanal, with occasional breaks to take in the magical and mystical mountains. These soothing vistas helped us regroup from the nightmarish weather in Chennai. Majestic eagles soared high up in the foggy skies as monkeys hopped from tree to tree, curiously gazing at us, hoping to get some bananas. But I knew not to feed them after the lessons I had learnt from Raj during our drive down the hills of Ooty. He had shared stories of many monkeys that got killed, trying to cut through the traffic to grab bananas from the drivers.

Over the next four days, Janani hosted just as many screenings, including in Chennai (again), Kodaikanal, and Coimbatore—my father's birthplace. This last screening was the most emotional to date. My father was front and centre in my thoughts. I was very aware of the hardships he'd gone through to make me the person I am today. He had never appreciated my work when he was alive. But he must have been smiling down from the heavens as I screened the film in the place of his birth. After Coimbatore, we conducted a few more impromptu screenings in the surrounding areas.

This included Sri Rangam, where a temple elephant named Aandaal, apparently, was forced to walk on three legs, based on some superstitious beliefs. We walked into the temple to face a massive crowd that had lined up at the entrance to get a glimpse of the deity. The ancient hand-sculpted rock carvings that graced the inside of the temple were exquisite, and so were the enormous statues that had also been carved out of a rock. The smells of the oil lamps and incense sticks were nostalgic as they reminded me of various pilgrimages I had taken with my family when I was a young girl.

Just as I was relishing my childhood memories, someone yelled at us. It was a priest wanting to know if I belonged to the Brahmin caste. In a rather open show of elitism, people of lower caste are not allowed inside the temple. It was another reminder that despite all the technological advancements India has made over the years, fundamental attitudes and mindsets were slow to change.

We looked for Aandaal, but she was nowhere to be seen. We asked around and discovered where she was. We peered into a tall structure through a tiny window on the backside. Here, we saw an elephant tethered to a pole inside a dark dungeon, standing on her urine and excrement. Adjacent to the structure was another shed. We knocked on it and a woman opened the door. She said her husband, Aandaal's handler, was not around and asked us to return in thirty minutes. We complied and returned with some fruits. But the handler initially refused to let us see the elephant, warning us against taking photos or videos. After a bit of convincing, he let us into her shed and allowed us to feed her the fruits we had bought for her. A short time later, we had to leave behind this lonely elephant in the dark dungeon, depressed and helpless.

After spending almost ten days with Janani, it was now time for us to part. Suddenly, my journey took a dramatic turn, and I would be asked to summon a whole new kind of resilience to weather the storms that would ensue.

Chapter 20

I Become a Shackled Elephant

Here I was making significant progress in raising awareness about the elephants' plight. Then the fateful day of 9 January arrived. On this day, I had visited the Kottoor Elephant Rehabilitation Centre, where I was feeding fruits to thirteen elephants after playing with two elephant calves. One special elephant at the camp, Rana, would change my life forever. As I approached him, he innocently opened his mouth. Into it, I placed a large piece of pineapple. However, he would not close his mouth. Assuming the pineapple was too large, I removed it, cut it into tiny pieces, and put it back in his mouth. I then fed him watermelon, bananas, papayas, and other fruits.

There was something about Rana's two white patches on his temples that caught my attention. I gently caressed the one on the right side. Then, looking into his gorgeous honey-brown eye, I said, 'Rana.' In that very moment, he suddenly butted his head against mine, the impact of which was so severe that it shot right down to my foot which was precariously positioned. My ankle twisted. I heard a few cracks. The bones collapsed and my foot swelled up, instantly rolling into a ball, the size of four hot cross buns on top of one another. I tried walking away from Rana, but the pain was so excruciating that I could barely move. My foot had twisted backwards. It quickly turned dark blue and was dangling. Everyone knew I was in trouble.

Meanwhile, the handler began to whip Rana. Yet, even as I was fading out of consciousness, I yelled out, 'Don't beat him! It's my fault! Please don't beat him!' One of the last things I remember is the sound of his shackles and his painful trumpets. The poor baby elephant was trying to escape the torture he was being subjected to by his handler, but he had no escape, for his shackles were too tight. I felt responsible for his suffering, I exacerbated his torture.

I was going in and out of consciousness. But with the help of five people who came rushing towards me, I was able to drag myself away. In minutes, they helped me into the backseat of a four-wheeler. The two women from the camp—Sita and Sajna—accompanied me. One of them held my head and the other, my foot. I was rushed to the hospital through the chaotic streets and dusty lanes. Finally, we arrived at the government hospital in Trivandrum.

Just as I was mending my broken relationships with my mother, with other human beings, and with my culture, I sustained severe fractures on my foot. I now had to mend my broken bones before everything else. Two metal plates and eighteen screws had been installed on my left ankle during surgery. They were holding my foot and ankle together until the bones could merge.

Back in my hotel room, while I was resting on my bed, the reality of what had happened hit home. I wondered what could have triggered Rana to act in this manner. Perhaps he was upset with me for having removed the pineapple from his mouth. Although I put it back after cutting the fruit into smaller pieces, maybe he kept his mouth wide open because he wanted more fruits. However, I had assumed that the piece was too large for a young elephant. Ignorance has no place when dealing with elephants. What a sin to have made the young elephant feel that I was teasing him with food. No wonder he became furious!

Later, someone told me, I was the seventh person that Rana had attacked. Had I known his history, I would have stayed away from him and respected the space he demanded. Clearly, he was in no mood to socialize with me after having been tormented all his life. Nor did he appreciate being touched. He hated being tethered even for a short while during his feeding time. These elephants were semi-wild and were released back in their enclosures after their meals. If and only if I had stayed with the two baby elephants—Arjun and Poorna—rather than having asked to feed the adult elephants, I probably would not have sustained a broken foot. I could have spent the whole day with them and observed how they were being looked after. Could have, should have, would have . . .

I was embarrassed to cancel a screening in Scotland organized by a passionate animal lover, Winnie. The two of us had been working on it since November 2016. Winnie had booked a prestigious theatre and generated posters and brochures. She had also connected me with Virginia McKenna, actress, author, and co-founder of the Born Free Foundation, who had received the Order of the British Empire, one of the highest awards in the United Kingdom. But I was in so much denial about the seriousness of my accident that until February, I was certain I could visit Scotland in March. What was I thinking? Eventually, when I came to grips with the magnitude of my injuries, I broke the bad news to Winnie. We had to cancel the screening. Winnie supported me unconditionally, despite the fact that she had taken so much trouble to plan the event.

More than two weeks had gone by since I'd had the surgery. It was now time to remove the sutures. Or to be exact, the metal staples that were embedded in my foot. On 19 January, I was escorted back to the hospital by Aathira, the young girl who bathed me regularly. We awaited the arrival of the orthopaedist in a congested passage packed with other patients in wheelchairs. I became extremely nervous. I dreaded the situation and wasn't at all certain how matters would turn out.

Ten minutes later, Dr Hari Kumar walked into the hallway. He said his 'hellos' while inspecting me, then gestured a nearby nurse to wheel me into his office. Once inside, the nurse soaked my cast in hot water. It disintegrated. He then began to peel the bandage away. My foot was still partially twisted. It dropped like a dead object as soon as the bloodied bandage was removed. After nineteen days, I was getting a glimpse of my pathetic left foot. I counted the staples. Fifty-six of them! At least, three staples in any given spot had been used to sew my skin together. This was mostly around my ankles on the inside and outside of my left foot. The skin was so dry that it resembled snake scales and had begun to flake. It was also now almost black due to a lack of movement and poor blood circulation.

Soon the doctor walked in with a twisted pair of scissors and forceps. He tried to distract me by striking up a mundane conversation. Then one by one, he cut each staple and plucked it out with the forceps as I screamed in excruciating pain. Tormented by my plight, Aathira left the room. After fifteen agonizing minutes, the doctor instructed the nurse to redress my foot. This time my cast would be a lighter one made of nylon fibre. After scheduling another appointment for 23 February, I left.

I had no idea where I gained the strength and endurance from to withstand the pain. Yet, my suffering paled in comparison to the agony and torture that elephants endure through no fault of theirs. They seldom receive the meticulous care and attention they deserve. These hapless animals are forced to stand on cement floors and walk on hot tar roads, despite the fact that their soft, cushioned feet are designed for lush jungles and marshes.

In captivity, their bodies are washed with detergent. But their skin is too sensitive for this. It's designed for mud baths, not chemicals. And I could barely imagine how the elephants must be coping with the weight of the shackles they wore. Chains weighing two hundred kilos are tied around their delicate ankles and tossed over their body. The metal digs into their flesh. Yet the poor elephants are forced to parade with raw, bleeding wounds. Many elephants suffer from arthritis.

Who knows how many of them had broken bones like me? Many are forced to parade limping, without even examining why they were unable to walk. But instead, the punishment becomes even more severe because they are unable to 'obey' their master's commands, not that they disobey purposely. You see, elephants cannot speak our language and do the best they can with what they understand. They cry out in agony, expressing their suffering, which is most often ignored by humans. Unable to bear the pain, they sometimes lash out and run amok. And if they do, they are tranquillized and brought under control using spike chains that dig into their flesh, then tortured for 'misbehaving'.

I thought of these elephants as I struggled to walk with the cast on my foot. It weighed down my hips, joints, and every single part of my body. On top of this, new discomforts were triggered daily. Strange sensations of thorns pricking or ants crawling up my leg and stinging my calf muscles. I made conscious efforts to keep my left leg in a normal position. But I was unable to do so. I was on painkillers, but their efficacy was short-lived. The pain I endured made sleep impossible. I could only lay on my back with my foot elevated on three pillows. My left hip and shoulders began to reveal signs of imbalance as chronic aches and shooting pains became the norm. I dreaded what the orthopaedic specialist would say during my next visit. I wondered what the X-ray would reveal. What if the plates had moved? What if the screws had shifted?

One night, after suffering relentless cacophonous questions in my head, I heard a soft, intuitive voice. It suggested that I had undergone this excruciating experience for a reason—by being incapacitated and forced to endure so much pain, I could feel what the temple elephants must feel.

All these weeks and months during the production, I had been able to relate to their emotional agony, but I needed to understand what they were enduring on a physical level. Could it be possible that I was chosen to embody the suffering of elephants so I could better articulate it? No matter what the reason for my suffering, I was in terrible pain. The only being that could help me was my spirit. It had always guided me through thick and thin, regardless of what my body and mind experienced.

Once again, I was reminded that despite my plight, at least, I had access to the outside world. I could connect with others via a phone call or social media. I had many visitors. I could also look outside my room at the mango tree. I could watch pigeons sitting on the windowsills chatting with each other. I could observe the birds flying freely. I could catch the sunrise through my glass windowpanes and feel the warmth of the sun's rays on me. I could write and release my feelings that way. But what about the elephants? How could they cope with their pain and suffering? People visit them, and some affectionately feed them fruit. This may bring them temporary joy, just as it did for me when people visited me. But when I was alone, the reality sank in. I was confined to my room just as the elephants were doomed to the small plot of ground they were shackled to.

In this state, I, too, experienced the physical and psychological pain and suffering that elephants endure in solitary confinement. Our physical and emotional sufferings were similar. The only difference was that I would be free someday. But the silent suffering of the captive elephants continues, and many are doomed to slavery until they die. Very few may be lucky enough to be confiscated by the government and kept in the government-run elephant camps. But even then, they would be unable to live out their true nature.

I knew now with certainty that I, on some level, had created the circumstance of my accident so that I could feel what the elephants feel. It was again proof positive to me that there is a valid reason for everything. This was a profound lesson I had learnt long ago, although soon after learning it, I had put it out of my mind. I tried to comfort myself by embodying this renewed wisdom, but still, I was unable to discard the feeling of hopelessness. When would this suffering end?

The same foot that had climbed hundreds of steps at the world-famous Sigiriya Lion Rock in Sri Lanka just a month prior was now unable to take even a single step. Would I ever climb the stairs again? Would I ever hike and walk the jungles again? Everything had been moving in the right direction, but suddenly, the brakes had been applied. Perhaps the universe wanted me to slow down. It had been over four years since the movie production had begun. I had never once stopped to smell the roses or celebrate my success. In fact, I had not taken a vacation in over a decade.

As if hammering home the point, on 30 January, the most unexpected disaster occurred. Aathira had just left the room after bathing me. But for no apparent reason, I called out to her and asked her to return. Suddenly, there was a power outage, and within a couple of minutes, we could smell smoke on the floor. In a matter of seconds, my room was filled with what appeared to be dense and dark clouds of smoulder, evidence of a serious fire. Try as I might, I couldn't open the glass screens. There were no windows in my room to release the smoke. We were now inhaling toxic fumes, and I was terrified.

Fortunately, within minutes, a maintenance worker rushed in. He unscrewed the glass panes and pushed them open. We had to stick out our heads to breathe in some fresh air. But the panic didn't end there. The smoke intensified and dense plumes hovered inside my room, and I began to choke. I thought I would die inhaling the noxious air. We heard loudspeaker announcements to vacate the building. People started scurrying down the stairs. But no matter how hard I hobbled with my broken foot, I could hardly move. Watching the helpless look on my face, Aathira tried to comfort me, but to no avail.

Suddenly, four men emerged out of nowhere. They announced that there was a fire on my floor, and I had to go down. Two more men brought in a wheelchair. They rapidly buckled me into it, and in minutes, the six men carried me down a flight of four stairs. There were precarious twists and turns and unpredictable steps. Grasping the handles of the wheelchair, I began hyperventilating. I was holding on for dear life, convinced that, somehow, I was headed for another fall. Terror struck! However, in a matter of seconds, we reached the ground level. Here, I was gently wheeled into the restaurant, where I sat for more than five hours. At around 10:00 p.m., after things had settled down, I was brought back to my room. The potent smell of the burnt wires still lingered, but the fumes had disappeared.

In all of this, I realized that during times of crisis, the only things that matter are compassion and kindness! The acts of unconditional service that the six men displayed helped to restore my faith in humanity. They reminded me of those who had rushed me to the hospital after my mishap at the Kottoor Elephant Rehabilitation Centre earlier in the month. Perhaps a deep awareness of the physical and psychological abuse that these gentle animals suffer could foster the same kind of love and compassion for elephants, too.

The very next day, I was presented with an opportunity to enlighten the police of Kerala about the elephants' plight. The state police chief (SPC), Mr Lokanath Behera, had arranged a wildlife symposium. I had also been in touch with one of the animal welfare lawyers. Hariraj Madhavan had attended my Thrissur film release and then organized the Kerala High Court show. He and I had been co-ordinating a presentation on the laws pertaining to wildlife and animal welfare, using a short video that we had extracted from the film. This, we would present to the police.

In further preparation for this, Hariraj arrived at 10:00 a.m. at my hotel on 31 January. Together, we reviewed the short videos on my computer and jotted down keynotes on the violations. Then, at 1:00 p.m. sharp, a car arrived at the hotel to take us to the Kerala Police Headquarters. The conference room was packed with high-ranking generals from all fourteen districts of Kerala, as ordered by the SPC, whose arrival we were eagerly awaiting. For the first time in my life, I was unable to go around the room and greet the attendees, as I was confined to my walker. Watching a waiter serve chai and spicy cashews made me realize the value of being able to walk freely. At around 2:00 p.m., the SPC and the chief conservator of forest walked in with a few guests. Everyone rose to their feet to greet them, and I struggled to join them. The chief gestured me to remain seated and took to the stage.

I settled in at the front desk that was outfitted with a microphone, with Hari seated right next to me. The chief then launched the symposium. He began by breaking the news that he was creating a wildlife crime unit (WCU) to end the atrocities against wildlife and captive elephants. In a matter of thirty minutes, the short film segment of *Gods in Shackles* was shown. There was total silence in the room. This was followed by my speech focused on the intrinsic value of Asian elephants. Hari utilized his legal expertise to explain how, in the first ten minutes of the short film, eighteen violations had been committed by the temple authorities, owners, and handlers. He then launched a casual discussion in the native language, challenging the officers to enforce the existing laws. The session ended at 5:00 p.m. Nothing could have stopped me from presenting a strong case to protect elephants at this very important conference that had taken seven months to become a reality.

For the first time in several years, the forest and police departments had come together under one roof to discuss the issue of wildlife crime in Kerala. The police had initially been apprehensive about what they would hear. But at the end of the day, they left with a better understanding of how the laws were being violated, and their role in enforcing them.

Despite having had a dramatic start, January 2017 ended on a positive note. Soon enough though, it was back to the task of healing. The newly constructed fibreglass cast outfitted on my legs felt good. But out of the blue, the plastic cups holding my heels began to cause throbbing aches and pains. The nylon material was rubbing against my skin. This caused severe itchiness, which, sadly, I couldn't scratch. During my bath, I poured boiling water into the cast to soothe the sensation. The doctor had convinced me that the water would drain off, but it didn't. Instead, it became lodged, and in a week's time, my foot began to emit a nasty odour. I dreaded the thought of foot rot and called the doctor immediately. But he asked me to monitor it for a week before visiting him.

Elephants also suffer from foot rot by standing on their urine and excrement. Here I was, paranoid of even the smell of a foot rot. What about the elephants? They suffer silently and are denied basic medical care. Once again, I couldn't help but contemplate the parallels that I was facing with my foot injury and those that the shackled elephants face all the time. In all of this, I marvelled at the Creator who could use me in this way.

These parallels brought to mind the adage: 'Love your neighbour as you love yourself.' In other words, you are them, and they are you. Never was this point made to me more poignantly than when I was suffering from my foot injury and the associated afflictions that accompanied it. During this phase, I couldn't help thinking about the countless captive elephants that suffered similar conditions every hour of every day—all over the world. The major difference is that while my injury was largely tended to and was expected to heal completely, theirs would never heal . . . Many die a miserable death due to neglect and abuse.

On 27 February, the doctors asked me to return to the hospital so that they could examine the itchiness. This time, Ashwathy, the hotel manager, accompanied me to the SP Fort Hospital. Fortunately, Dr Cherian was in attendance. He immediately asked the nurse to slice off the nylon cast, four weeks ahead of schedule. I was taken into a special room where the nurse brought in an electrical saw which he plugged into a wall socket. The moment he placed it on the cast and began to drill, once again, terror and panic imbued my body. I was convinced that the saw would slice off my bones. It was one of the most terrifying thirty minutes of my life. But thankfully, everything went smoothly enough.

The cast was now finally removed. And for the first time in eight weeks, I could touch the skin of my swollen left leg and scratch all I wanted. As I caressed my foot, I sensed numbness near my ankle. But the doctor said that it would take a few weeks for the sensation to come back, with proper blood circulation. I was then given an air walker. This was a kind of removable cast, a heavy boot with a massive bottom to support my ankle and toes. I could now nurture my foot and the encrusted skin, knowing that the physical structures would adapt. I began to cherish simple pleasures like bathing. I enjoyed the primal sensations of scratching and touching. The weight had been shed and my foot was much lighter. But now my body would have to learn to rebalance itself.

It would be five more months of rigorous physio and massage therapies before I would be able to walk unassisted. At that point, I would work to eliminate my slight limp. I was lucky to be this far along in the course of my healing. And I promised myself that I would never take my body for granted again.

I was ready to return to Toronto as I was feeling homesick. I missed my long walks in the wilderness, the sweet melodies of the birds, the snow, and my independent and solitary life. Normally, I would start my day by communing with the natural world. Somehow, the mornings felt deeply mystical in the woods. The breeze, the sweet melodies of the birds, the bubbling brook, and the rustling sounds of trees soothed my body, mind, and spirit. Rain, snow, or shine, Mother Nature's energy always recharged my own vitals. I was now feeling desperate to place my naked feet on the ground and absorb the earth's soothing energy.

Back in the hotel room, I tried to get a dose of Mother Nature by looking out my glass windowpanes. Every morning, I eagerly awaited the emergence of the magnificent urban sun and revelled in the resplendent eagles soaring toward the heavens. Watching the cranes, herons, and ravens deftly flying high against a backdrop of clear blue skies was liberating. Every evening at 6:00 p.m., I anxiously awaited the arrival of the flock of egrets that lived in the mango tree outside my room. At least, fifty of them congregated and socialized before they nestled into the branches to retire for the night. I would then draw my curtains and retreat into my own bed. Mother Nature nurtured me and played an important role in my healing. As I was feeling hopeful and bouncing back during the bleakest February of my life, the greatest surprise of the year would arrive the following week.

Chapter 21

Nari Shakti Puraskar

It was the afternoon of 23 February when my phone rang, with a New Delhi number flashing on the screen. A faint voice on the other end wanted to confirm whether or not she was speaking with Sangita. Then, Ridhi disclosed that I had been selected for the Nari Shakti Puraskar, which translates to Woman Power Award. This is India's highest distinction given to women who are making a difference in India. She said that the president of India would be handing out the awards to the select thirty-three women on the occasion of the 2017 International Women's Day. I was being conferred the award for exposing the atrocities against elephants through my documentary.

I was utterly dumbfounded! Choking back my emotions, I explained that I was physically unfit to travel as I was recovering from a serious accident. However, if my presence was absolutely required, I would be there.

I turned to the concierge manager, Ashwathy, and asked if Neethu, one of the other young women who'd been taking care of me, could accompany me to New Delhi. The permission was instantly granted. My physiotherapy had begun just two days before I flew out to New Delhi, provided by the same doctor who'd performed routine check-ups on me when I was hospitalized. The swelling had subsided by fifty per cent in two days.

Neethu showed up and patiently helped me pack my things, despite my occasional outbursts with her. In that very moment, I realized that whenever I became impatient, the tension travelled to my foot and stiffened the muscles. It dawned on me then that I was actually becoming impatient with the pace of my healing. I had to constantly remind myself that I could not push against the current. I had no reason to become impatient, for everyone was doing their best. Among other things, the perfect physiotherapist had been placed on my path at the perfect time. And the perfect caretaker was accompanying me to the most memorable event of my life. My only job was to allow my body to heal and to be compassionate towards myself and everyone I came in contact with. Joseph Campbell once said, 'Our life evokes our character. You find out more about yourself as you go on. That's why it's good to be able to put yourself in situations that will evoke your higher nature rather than your lower.' It seemed I was attracting compassionate people who were drawing out my higher nature.

In the ensuing weeks, the media contacted me to feature my story on the national news. Given my physical limitations, the Union Ministry of Women and Child Development (UMWCD) had arranged a TV interview at my hotel in Kerala. As Ashwathy escorted me to the room where the interview had been arranged, I had flashbacks of my nine-year tenure as a journalist where I had to dig out stories and research my subjects. It was a sweet feeling to realize that the tables had turned; now, I was the subject and the story.

We arrived in New Delhi early the next morning and were driven to the hotel. As we stepped outside the car, a brisk northerly wind brushed my face, reminding me that winter was still lingering. A waiter wheeled me in a wheelchair, with Neethu following us to a cozy room at the Samrat Hotel. After a light breakfast and a cup of chai, it was time to rest and recoup from the overnight travel fatigue.

Later that evening, there was a small gathering in that

same hotel, where other award recipients were mixing and mingling. This group of brilliant women consisted of authors, calligraphy artists, naturalists, conservationists, policewomen, businesswomen, and even a gorgeous and renowned actress. It also included a young entrepreneur, Zuboni, from Nagaland, as well as a woman who had climbed Mount Everest. These women, from every walk of life, were not only incredibly talented and accomplished, but also humble.

One of the award recipients was an acid-attack victim featured in the Bollywood actor Amir Khan's television series, *Satyamev Jayate* (Truth Alone Prevails). It felt surreal to meet a woman who was brutalized by her husband for dowry, a stark reminder of the persistent female subjugation in a patriarchal society. She was an inspiration to all the battered women of India. It was also deeply inspiring to meet many wildlife warriors—a popular south Indian actress, Amala Akinneni; wildlife conservationist, Pamela Malhotra; and naturalist, Dr Nandita Shah—who had all achieved major milestones in nature conservation.

The grandest day of my life soon arrived! I was all decked out in my dark green kurta (long knee-length shirt) and tights, waiting for the other women. All of them showed up with beautifully painted toes, wearing glamorous sandals. Sadly, my own left foot was wrapped in grey medical boots, and my right foot was adorned in a semi-glamorous sandal. I took a deep breath, trying not to get too caught up in vanity and external appearances.

At the country's highest house, a gentlewoman named Swati helped me through the entrance and through the hall into the Rashtrapati Bhavan—India's White House. We arrived early for a dress rehearsal of sorts before the award ceremony itself. Shweta, my constant companion at the ceremony, remained by my side throughout, providing me with strength and support.

I can't even begin to describe the feelings that went through my mind as I was wheeled into this historic building. Looking at the paintings of great leaders who had fought for India's independence from the British, gave me goosebumps. The brave faces of Subhash Chandra Bose, Chandragupta Maurya, Mahatma Gandhi, Jawaharlal Nehru, and others stirred up childhood memories. I had read about them in history lessons during my primary school years. But to visit the same building where these great leaders had convened felt like an illusion. I closed my eyes in silence and paid my gratitude for the freedoms I enjoy because of their sacrifices. I also asked their souls to guide me in my quest to liberate the captive elephants, who continue to suffer the fate that Indian people did during the British regime.

I was then wheeled through the hallways and into the elevator. Glancing at the extraordinary ceilings that portrayed India's culture and history, my heart flooded with nationalist emotions. The main hall where the ceremony was to take place was packed with high-ranking military admirals who served India's president. They led us through the practice run for about an hour. Shortly thereafter, we were asked to rise to our feet as the band played India's national anthem. The anthem serenaded the Honourable President, Pranab Mukherjee, who was then escorted to his throne. I suddenly felt incredibly patriotic, falling in love all over again with the same culture that I had loathed for decades. The barriers and past resentments were melting, and I was beginning to heal.

Then came the relentless announcements and live coverage on the state television station—Doordarshan. One by one, all thirty-three women were recognized. Extraordinary efforts had been made to ensure I was well cared for. One of the military admirals held my hand and escorted me to the president's stage. The President then climbed down the steps. Looking me directly in the eye, he handed me my Nari Shakti Puraskar, an award that I received on behalf of the elephants. It was a truly special moment. Later, we took a group photo with the president. It was also a great privilege to be surrounded by such wonderful, dynamic women who had accomplished so much with their lives and empowered to serve their grander purpose by following their heart's true calling. They were incredibly inspiring.

That night, Minister Maneka Gandhi had organized a special dinner celebration for us. I had seen this woman many times on television and in person. This included the screening in New Delhi in June 2016, and when I was walking the red carpet to receive my award from India's President. But to sit at her table and listen to her stories was truly heart-rending. She spoke bluntly about the lack of animal protection in India, the burgeoning problem of human population growth, and how her husband had been rebuked by even Western authors, who called him an authoritarian when he had tried to implement birth control on a nationwide scale. It was also amazing to socialize with Ridhi, the woman who had initially called me to inform that I was one of the award recipients. Then there was Shipra, who directed the event. These humble women cheered us on while receiving our award.

After supper, six women came to my room, where we talked until the wee hours of the night. Zuboni from Nagaland acknowledged that people in her state consumed elephant meat. This stirred up a passionate but friendly debate between the vegetarians and the meat eaters present. Zuboni was not a bad person, and she told us that she had never eaten elephant meat. That said, she hadn't spoken out against it either, which prompted me to remind her that if good people don't stand up for the truth, bad things will happen. People in Nagaland may not be bad, yet they do ignorant things, given their lack of knowledge about the intrinsic value of the elephants.

The next day was even more exciting. The award recipients were scheduled to meet the prime minister, the

Honourable Narendra Modi. It may sound vain, but for the first time in three months, I went to the salon to pamper myself. After getting ready, Neethu and I arrived at the hotel reception just in time to be picked up in a special car. We were then driven to an unknown location. It turned out to be the private residence of India's prime minister. Security was tight, and after going through a metal detector, we emerged into a lush green courtyard, fenced with tall cement walls. Hibiscus, roses, jasmine, and other endemic flowers were in full bloom. As I walked with the support of my walker, a whiff of warm breeze carried the sweet scent of these exotic flowers. It was hard to believe that I was inside the private house of India's prime minister, the most powerful man in the nation, the captain, steering the country. His powerful, yet compassionate words drew millions of people to his rallies. And PM Modi had risen to this significant position just two years prior to this gathering.

We were escorted into a conference room with about fifty chairs arranged in theatrical style and a row of about eight to ten chairs across. In a few minutes, local members of parliament began to arrive, followed by the minister in charge of these celebrations, Maneka Gandhi herself. At around 3:00 p.m., Mr Modi arrived through the back door, bringing with him his magnetic aura and a powerful presence. Everyone rose to their feet.

Just as he began to address us, the heavens opened up. Soon, thunder rolled over the roof. Rain is considered a good omen in India. The prime minister paused for a moment as though welcoming us and Mother Nature's blessings. He then ordered the chief secretary to introduce us. After this, he gave a short and heartfelt speech. And in a surprise move, he mingled with us for a few minutes, allowing selfies with the young women. He suddenly turned to me and asked what had happened to my foot. He had clearly taken note of my physical condition. After explaining what had happened to me, I pleaded with him to save the elephants, using the opportunity to be a voice for the elephants.

As we stepped outside the conference room, the smell of rain emanated from the earth and the sun broke through the clouds. The verandah had become slippery and I had to be wheeled out in order to avoid another accident. That evening, everyone gathered in the hotel restaurant to dine together one last time.

My mother phoned me the next day, stating that my brother had shared my photos with everyone. Apparently, the entire family was proud of me. She told me that after the live coverage, her phone was bombarded with calls. Various relatives called her about the good reputation I'd garnered for the family. It made me wonder if they were the same relatives who had washed their hands of me when I'd embarked on my unconventional career. I'm certainly glad that no matter what anyone said or did, I'd followed my heart. In so doing, I'd been led to undertake my life's work, learning so many rich and life-transforming lessons along the way.

Around this time, a UK actor named Dan Richardson, with whom I had connected with on social media, had reached out to me, in what would turn out to be another synchronicity. I had commented on a video I'd seen of him protesting against the captivity of dolphins in Japan. We Skyped the same week. Our first conversation lasted for more than two hours, and I learnt that he was an ambassador for the renowned Born Free Foundation, a wildlife conservation organization based in the United Kingdom. He had also travelled to Africa to speak out against the ivory trade in China and the perilous situation of the rhinos. One thing led to the next, and he offered to screen *Gods in Shackles* in London in October 2017, marking the European debut of the film. The screening catapulted a subsequent peace protest in front of the Indian embassy during Mr Modi's visit, demanding the prime minister to ban elephants from cultural festivals.

I now had only twenty days left in India, and there were

many loose ends for me to tie up in Kerala. Despite my daily physical therapy, my entire body was out of balance. And I still had to meet the Swamiji who was the darling of my mother. He could be a great ally in helping end the atrocities against the elephants. So, I phoned the ashram and booked a time to visit Thrissur on 25 March. This was just days before I was due to fly back out to Toronto.

In the meantime, I organized a groundbreaking meeting between the chief secretary of Kerala and the world-renowned elephant scientist, Dr Raman Sukumar, featured in my film. Dr Sukumar made a day trip from Bangalore just for this one meeting on 16 March. After the discussion, Sukumar and I spent an entire day together, catching up on life since the movie's release in July 2016, while also discussing potential collaborations in the future. The same scientist who was apprehensive about giving me an interview the first time had now become one of my closest allies.

After dropping him off at the airport that evening, I made another presentation for the outgoing chief wildlife warden, the head of forest force, and the chief secretary, who bridged the gap between the forest department and me. This high-level meeting opened up doors to provide a symposium for the forest department on 22 March.

Two days later, I took another bold step to travel to Thrissur. This was after I'd discarded my walker to the physiotherapist—but I took comfort in knowing that I had Neethu to lean on. Along with heavy police protection and the same personal security officer (PSO), we travelled to the Narayana Tapovanam Ashram after midnight, trying to avoid the traffic. As soon as we entered Thrissur district, a police jeep escorted us to the countryside. It was peppered with rich estates interspersed between patches of lush forest. The sun was coming up and its rays danced through the mango and jackfruit trees, with the sounds of exotic birds echoing through the air. The village was just waking up, the sounds of temple bells and traditional songs blaring through the streets. In thirty minutes, we made it to the ashram's entrance. Seven women in white saris with orange borders and some sandalwood paste on their foreheads were standing on the verandah.

By now, my foot was swollen and throbbing, as it had not been elevated in the car during our long drive. The first thing Neethu did was massage it with a special cream to allow blood circulation. My room faced a garden where white, red, and yellow flowers were blossoming. Butterflies hovered over the blossoms, displaying their colourful wings. The sweet scent of ripe mangoes tantalized my senses. The sun had intensified making the stagnant air feel like that of a furnace.

I walked down the steep steps gingerly, holding onto the precious walking stick that had replaced my walker. And after a delicious breakfast, we waited in the library for about thirty minutes before I was approached by a young woman, the right-hand person to the spiritual leader. She told me that Swamiji was on the second floor and that she would escort us there. As though she knew what I was thinking, she said, 'Don't worry, we have an elevator.' The hallways felt very maze-like and endless. Finally, I was face to face with the same holy man I had seen on television at my mother's home, wearing an orange sarong and shirt. In his late eighties, Swami Boomananda Tirtha had a special glow in his eyes.

I teared up and told him that my mother had inspired me to visit him, and that she and I had religiously watched *Spiritual Splendour* during my visit to Mumbai. He held my face with his palms, as though holding the face of a little girl, and looked into my eyes, with kindness and compassion oozing through his wise eyes. I touched his feet, honouring my traditions and handed him a plate of fruits and dhakshina (donation). I also handed him a copy of my documentary, *Gods in Shackles*.

After we had settled down, he began to talk passionately about the plight of elephants. He explained that he had met key temple authorities in his attempts to persuade them to discontinue the use of these animals in rituals. But his requests were vehemently rejected. We discussed a few solutions that I'm unable to share at this time. It occurred to me that mending my relationship with my mother was helping me to reconnect with and heal my cultural roots, which is absolutely critical to ending the atrocities against elephants.

Suddenly, the journalist in me woke up, and I felt compelled to interview this spiritual leader using my iPhone. One of the most disturbing things he revealed was that elephants were being fed meat on a regular basis. Elephants are vegetarians, and their guts are not designed to digest meat. They are enormous and need to consume 200 to 300 pounds of a wide variety of vegetation daily to meet their nutritional needs. I began to connect the dots. Perhaps this could be why hundreds of captive elephants were dying of digestive disorders in Kerala. I then wrote a report about this disclosure in the popular online portal *Huffington Post*, which, not surprisingly, stirred controversy. Temple officials visited the spiritual leader's ashram and threatened to sever all ties with him. It must also be noted that this same leader had filed a Supreme Court case against the exploitation of elephants. As of July 2020, a verdict is still pending.

On 11 March 2018, the news broke out that Shiva Sundar, the star elephant of Thrissur Pooram, who was also featured in my film, had died of stomach impaction. A massive ball of fibre had clogged up his digestive system. The poor elephant was constantly fed cheap fibrous Caryota palm leaves, denied adequate water and the variety of fodder that elephants require to meet their basic nutritional demands. Furthermore, being a star elephant, he was forced to parade beneath the scorching sun at as many festivals as possible. He fetched hundreds of thousands of dollars during the festival season that takes place between December and May every year.

As Swamiji and I were exchanging our encounters with Shiva Sundar and grieving his death, I realized that

I had to prepare for another screening that night for the police cadets at the Kerala Police Academy in Thrissur. We drove to the police quarters where *Gods in Shackles* was screened to the uniformed men. Soon after the movie ended, there was an open discussion. In this, I had a chance to interact with some passionate environmentalists in the police department. I left knowing that I had done my part to heighten awareness on the plight of elephants. And it seemed to apparently have resonated with the tough police guys, who vowed to do everything they can to protect the elephants of Kerala's cultural hub.

We arrived in Trivandrum early the next morning. With just three days left before leaving for Toronto, I had a very important visit to make. I needed to commune with elephant Rana, so that we both could heal from our disastrous encounter. I absolutely had to apologize for the inconvenience I'd caused by taking the pineapple out of his mouth and creating panic. I carried with me kilos of fruit. This time, I allowed the handlers to feed the elephants, observing Rana from a distance. As I kept repeating silently, 'I love you, Rana, please forgive me', I could feel my tears dripping on my blouse. Meanwhile, Rana devoured the pineapple, and I was savouring every moment admiring this sweet young elephant and feeling a profound sense of oneness. We were healing!

When I arrived at the Kottoor centre, the two baby elephants, Arjun and Poorna, were resting inside their quarters. As soon as I peeked inside through the windows, the adorable little ones stuck their trunks out through the bars. I was then driven around to see other elephants. I thanked the handlers and the forest staff and promised that I would return to work with them. The following day, I hastily arranged meetings to say my final goodbyes to the retiring chief secretary and the additional secretary of the forest. They had paved the way for forest department screenings and provided access to the state's largest elephant camp.

I was originally scheduled to return to Toronto on 31

March in economy class. But given that the seats tend to be congested and cumbersome, I was terrified that the swelling of my foot would be exacerbated. So, I tried to upgrade my ticket to first class, but that would cost me three thousand dollars—precious funds that I would rather save up for the elephants. Around the same time, a generous supporter named Lori Valdez contacted me. She offered to fly me back in business class at a much cheaper price as she worked for a renowned airline. Once again, the universe provided for me.

And as fate would have it, I had to fly out of Mumbai for this flight. Thus, I got to spend three more days with my mother. I realized how much things had changed since the happy times in December 2016! Although she tried to mask her emotions, I could tell that she was shocked to see my shackled foot. On the other hand, she was also very excited to see the photos and videos of her guru. She talked about it endlessly with everyone who visited and phoned her. Despite her ailing health, she had taken the trouble to make delicious almond squares, as she knew how much I loved sweets. We sat at the same portable table and dined together. We laughed, joked, and reminisced. The flow of her unconditional love continued on from my December trip. We had certainly mended our broken relationship. I had to now return to Canada and continue to mend my broken bones.

Meanwhile, Lori stayed in touch with me, keeping me informed about available flights. At one point, I said to her, 'I have been put through so many tests, through such uncertainties. My faith and trust in the universal powers are constantly being tested. And they only become stronger. Thanks to angels like you who have been there to support me. You are of course God-sent!' Lori responded by saying, 'I totally believe that sentiment. It is a story for another day, but God spoke to me loud and clear. Three years ago, he commanded me to help the elephants. I did not know how, but He has laid the path and today I'm following it.'

On the day of my departure, my brother took me to the

airport. Here, he also secured a wheelchair that Lori had arranged for me. In a matter of minutes, I was wheeled away. I waved goodbye to my brother until he disappeared into the crowd. I was taken to the first-class lounge, where I had a couple of glasses of my favourite sparkling wine. When it was time, a waiter wheeled me into the plane and dropped me off at my seat where a cuddly stuffed white bear awaited my arrival. I hugged him all night like a baby; he gave me solace.

I arrived in Toronto on 4 April. Spring had sprung and so had my hopes for healing. For the next six months, I underwent painstaking physical and massage therapy, three times a week. And not surprisingly, the most perfect therapists were placed in my path. Dr Bahram Jam worked on strengthening my bones using rigorous exercises. Ling Turco, my massage therapist, worked on my muscles through muscular manipulations. And one of the most healing therapies to release my painful memories was provided by Amy Chau.

Using brain exercises, Amy literally exorcized some of the audiovisual memories stored in specific muscles of my body. The sounds of Rana's shackles when he was whipped after he butted my head were stored in the form of guilt and shame in my shoulders. I was also shouldering the sounds of my walker. It had certainly supported me, but on a deeper level, it had left me feeling that I would be disabled forever. Three elephants that were featured in my film had died as a result of the abuse they'd suffered. But unfortunately, I was too helpless to do anything for them. Shiva Sundar, the star elephant, Ayyappan, and Ramabadhran were now liberated.

As I grieve their loss even today, every teardrop I shed is giving me emotional freedom. It is now my turn to liberate my soul animals, and I am committed to doing everything in my capacity to unshackle the elephants that have helped me with liberation from my own suffering.

Chapter 22

Full Circle

The year was 1989 when I had fled to Canada. I never wanted to return to India, for my creative spirit had been crushed by what seemed like an oppressive culture and a patriarchal society. I had abandoned the traditions that had shackled my mind and suppressed my innate potential, preventing me from soaring. The freedom in the western world allowed me to identify my dysfunctional mental conditioning. It emboldened me to challenge the cultural myths I'd grown up with while also transforming my world views and perspectives.

Witnessing the atrocities against elephants reminded me of my own plight. When I embarked on my four-year journey to produce the movie, anger and resentment towards those who inflicted suffering on the hapless elephants behind the veil of culture had been intensified. But by the end of those four years, my perspectives and perceptions had changed dramatically. The awards, accolades, and acclaim stemming from my movie certainly offered a pulpit to expose the relentless neglect and abuse of elephants. However, something much grander had transpired. I struggle to come to grips with it even today as I try to summarize my life-changing odyssey with the shackled elephants.

So many unthinkable circumstances had unfolded, and strangers had been placed on my path to help bring to light the stories of the suffering elephants. What if, deep down, people wanted to right the wrong? What if, even unwittingly, people wanted to end the injustices against these majestic animals? Perhaps social norms and cultural shackles were preventing them from doing what they knew they should do. But no matter what, I had identified my role of a messenger, carrying forward whatever the elephants wanted me to convey. In the process, I went through the darkest nights of my soul. I met people who pushed all the wrong buttons and triggered my shadow. I cried myself to sleep several nights, mourning the deaths of so many elephants. I wanted to give up so many times, but I simply could not turn my back on the elephants after having witnessed so many atrocities against them.

Every step of the way, I journeyed into the wilderness where I spent uninterrupted solitude. Here, Mother Nature's creatures reminded me to simply be and allow the life's flow. This helped me to connect with my true self. It gave me the freedom to discover my real identity and true purpose without anyone trying to tell me what I was or wasn't capable of doing. After mapping out my journey, I did whatever gave me a sense of contentment. Did I struggle with fear and trepidation? You bet I did! It happened almost every single day. And yet the natural world showed me how to transcend my own subjective fears and do whatever I needed to do for the elephants, thereby serving the grander good.

Along the way, I met kind and compassionate people who made me realize that there is goodness in this world. They do the best they can with what they know and have. Culture has shaped their beliefs and behaviours to a great extent. In this, they have been unwittingly brainwashed by misguided myths.

Animals, in particular, are vulnerable to these false beliefs. They cannot speak human language, and most people do not understand or tend to ignore the language of animals. Therefore, the onus is upon us, humans, to shift our attitudes and behaviours that inflict any kind of suffering on any beings—people, animals, critters, or trees. This would require an open mind and willingness to change, which is only possible through awareness and empathy.

No doubt, the people of Kerala had demonstrated compassion and kindness towards me when I was hospitalized. They helped the process of mending my broken bones by shining the goodness of their true nature and treating me as their own. This clearly demonstrates that humans intrinsically yearn to help those who suffer, even a stranger. Surely, we can find a way to transfer the same kind of caring and compassion towards those who don't look like us.

This is vital to healing the planet and all living beings. Unity is crucial to healing the elephants, for this is a very critical time in our planet's history. Opening up the hearts and minds of people to practice ahimsa or non-violence—which originated in India—can foster love, compassion, and kindness.

India's own legendary Mahatma Gandhi's Civil Disobedience Movement, which employed the construct of non-violence, is a classic example of ahimsa. When he faced discrimination in South Africa and was pushed out of the first-class compartment violently, he refused to retaliate. Through several peaceful marches, including the famous Dandi March in India in 1930, this supreme soul manifested India's freedom from the British. So many of the world's renowned leaders—that include Mother Teresa, Martin Luther King Jr, and Abraham Lincoln—fought oppression and slavery by extending love and compassion.

In a similar way, to end elephant slavery, we ourselves must embrace ahimsa. As hard as it may seem, we must do our best to empathize with even those who inflict suffering on elephants. We need to enlighten people with knowledge so the darkness of ignorance may be removed. Violent thoughts, words, and actions will only perpetuate the suffering of humans, which will spill over to the elephants and keep them enslaved until they die.

From hatred to compassion to empathy, I have come full circle, having accepted myself and my frailties. From loathing a culture that put me through so many miseries, today, I feel a sense of belonging to it. And now that I have seen the brighter side of my culture, I have embraced it and healed my fractured relationships as well. I sure am grateful to the elephants for having shown me the way and creating memorable encounters, not only with their species, but also with human species.

23 June 2016 was a very poignant day for me. It was the day I landed in Kerala, hoping to release the film there. It was also the third anniversary of my father's death. My whole journey, that of making *Gods in Shackles,* had begun on the first anniversary of his death, in June 2013. I don't believe that the way the dates aligned as bookends can be put down to just a coincidence.

My entire journey into exposing the plight of elephants has been intuitively guided. It had been designed meticulously by the Creator, to unshackle my own myths and mental conditioning so I can then work with people to unshackle the suffering elephants. Providence paved the way and led me along a path that began as a calling. This same path offered me rich and profound perspectives on life and other people along the way. Even strangers ended up inspiring me when I allowed them to be who they really were. For they, in turn, elicited my authentic spirit.

There are still deeper layers of emotions that need to be peeled off. But it is much harder because the deeper the emotions and the more hardened they become, the more difficult they are to unshackle and release. For now, my faith in humanity has been restored. I can harness the relationships that I'm building to change attitudes and mindsets.

By exposing the suffering of elephants, my most sincere intention is to help humanity become aware of its man-made cultural shackles. These shackles inflict pain and suffering on our planet's second-largest mammal, one of the most conscious and compassionate animals on earth—the Asian elephants. This species is being pushed to the brink of extinction due to human activities driven by greed, selfishness, and cultural myths.

Since human beings have created these problems for elephants, humans are the only ones who can implement existing solutions. At the same time, humans can work towards creating even better ones, while harnessing their innate potential to serve the greater good of the planet and all of its inhabitants. In so doing, they will also benefit on so many levels. Human beings are intrinsically connected by a common purpose and one universal spirit that longs to be united with each other and all living beings.

The creatures of the earth live harmoniously by expressing their unique gifts and nature. They teach us to simply be and accept our self for who we are. In doing so, we can accept others for who they are and project compassion onto the outside world. Contrary to the myths instilled in our mind—that we have to sacrifice our well-being to pursue our purpose—when we retreat into our inner world, we can heal our self as frequently as necessary. This, in turn, helps us to return to the physical world with clarity and a new meaning to our lives.

We are intricately connected to this magnificent web of life. We are tributaries of one ocean, and we are blood vessels and organs of one body—the planet Earth. If any of these components becomes damaged, the entire body will collapse. We need to come together to heal the planet and the elephants. If humans can collectively unleash the shackles that confine them—the shackles of culture, material wealth, and status quo, or whatever they may be—we can become compassionate enough to heal all sentient beings, as well as our own pain and suffering. But first, we need to heal our self by reconnecting with our origins—the wilderness and its inhabitants, so that we can foster a peaceful coexistence.

Acknowledgements

I am profoundly grateful to elephant Lakshmi for her unconditional love and trust in me, despite having been brutalized and broken by my species. She evoked deeply buried feelings about my own subjugation and abuse by exposing the assault and brutality that she had endured. Not a day goes by that I don't think of my soulmate. Thank you, Lakshmi, for accepting me for who I am, and for welcoming me into your space.

My sincere apologies to elephant Ayyappan for the pathetic life and suffering imposed on him by my species. Although I exposed your suffering, I regret that in a short few months you crossed the rainbow bridge, leaving me with no opportunity to rescue you. By spending some time in solitude with you and watching how your caretakers totally ignored you, depriving you of the basic necessities, my heart shattered into a million pieces. But your suffering shed some light on how I had been depriving my soul's basic necessities. Amid all your suffering, you also expressed your playfulness during the brief time I spent filming you. It lightened me up—just a bit!

I am grateful to elephant Shiva Sundar for recognizing my genuine love and reciprocating his pure love, even as he was just about to be abused. It seemed like I was placed at the scene in time to dissolve the tension. I can still hear you release a heavy sigh of relief. That moment will remain etched in my mind until I die. I am grateful to have had the opportunity to spend some time and share my love with you.

I am so deeply grateful to elephant Ramabadhran for

mirroring my own emotional paralysis by dangling his paralyzed trunk. I realized how my mental paralysis was preventing me from moving forward. His heartbreaking suffering caused by fear—an emotion that paralyzes humans from thinking loving thoughts—reminds me of how much work is ahead of us to heal the emotional wounds of humans.

I appreciate bull elephant Jairam who revealed to me how fear turns to cruelty. As of September 2020, he is still a prisoner, shackled and forced to perform in the name of culture and religion. It has been challenging finding a way to unshackle this living God—Lord Ganesha. I am grateful to Jairam for letting me hose and scrub him, and for the few moments of intimacy we shared together.

Mother Nature continues to nurture me. I am grateful for the deep connection I share with Her and my non-human brothers and sisters. By communing with Her for hours during my writing journey, She offered me deeper insights. The geese family was among the creatures of the earth that comforted me. When the goslings hatched in the early spring of 2016, it was a very poignant moment that I have elaborated in a chapter.

I am profoundly grateful to the universe for aligning human animals on my path, some of whom inspired and encouraged me, while others triggered my inner demons. I am grateful to all of them, especially Sukesan, Sita, and Sajna from the Kerala forest department who restored my faith in humanity. I will always remain grateful for their unconditional care and service when I most needed it. Every individual played a significant role in making me a better person and I am deeply thankful to them.

I am eternally grateful to Dr Jane Goodall, an authentic soul, a heart-centred conservationist, and a humanitarian, for being part of my life's journey. I am deeply honoured, and still have to pinch myself to believe that despite her hectic schedule, she took out precious time not only to review the book, but also write a heartfelt foreword.

I also appreciate Richard Louv, a prolific author and creator of the term, Nature Deficit Disorder (NDD), whom I had met in Toronto through the Royal Roads University where I obtained my Master's degree. Ten years later, and after following my journey, I am honoured that Richard, an author of numerous books about nature, has also written the foreword.

I am indebted to Deena Metzger for providing me with meticulous and valuable feedback. Her reflections evoked deeper insights that I could incorporate in my book. I am also deeply grateful to Deena for endorsing my book. And to Rula Lenska, Margrit Coates, Dr Liza Ireland, Dr Marc Beckoff, and Carla Kovach, thank you for taking the time to read my book and offer your thoughtful reviews.

I would like to thank my editor turned friend, Anne Dillon, a kindred spirit helping wildlife in her own way, for providing unconditional love and support throughout my writing journey. I am also grateful to KN Literary for matching me up with Anne.

I appreciate Ruchita Ahuja for affording me the opportunity to work with this world-renowned publishing house, and the entire team at Hay House, India, for their tremendous support and input.

I appreciate my family for their love and support of my mission, especially my late father, whose spirit I believe watched over my journey every step of the way, guiding me to the perfect destinations, humans, and non-humans. I honour my parents and grandparents for instilling Hindu philosophies and values of hard work, Dharma, and Karma, and for constantly reminding me of the ripples that our thoughts and actions create and their impact, not only on ourselves, but also on all living beings.

Above all, I revere the Spirit that dwells inside me—the same Spirit that dwells in you—a higher power that shines through each ray of the magnificent sun, every sand grain on the shores of the ocean, every wave lashing on the shore, every twinkling star, and every fluff of the clouds in the sky. The same Spirit that speaks to us through every butterfly, bee, ant, snail, and every creature small and large. This gentle Spirit weaves together all sentient beings, creating a profound sense of oneness with humans and non-humans that continue to cross my path.

Author Bio

Sangita Iyer is a National Geographic Explorer, multi-award winning nature and wildlife filmmaker, broadcast journalist and biologist. She is also the executive director and producer of the globally acclaimed epic documentary, *Gods in Shackles*, which was nominated at the United Nations General Assembly and has garnered a dozen international film festival awards. The film exposes the atrocities against captive elephants that are exploited for profit behind the veil of culture and religion, and has inspired strict enforcement of captive elephant management rules in the southern Indian state of Kerala.

She received the Nari Shakti Puraskar (Women Power Award) – the highest award for women making a difference in India from the country's president – for her courage to expose the plight of captive elephants. Sangita's short documentary series, *Asian Elephants 101*, that she produced, directed and hosted, has aired on multiple documentary channels. She produced the short films after receiving a storytelling award from National Geographic Society. Sangita is also the founder of Voice for Asian Elephants Society, an elephant conservation non-profit organization with a mission to protect India's elephants.

Sangi and Poorna

HAY HOUSE

Look within

Join the conversation about latest products,
events, exclusive offers and more.

Hay House

@HayHouseUK

@hayhouseuk

We'd love to hear from you!